T0296303

Energy Sustainability

Energy Sustainability

Ibrahim Dincer
University of Ontario Institute of Technology
Oshawa, ON, Canada

Azzam Abu-Rayash
University of Ontario Institute of Technology
Oshawa, ON, Canada

Elsevier
Radarweg 29, PO Box 211, 1000 AE Amsterdam, Netherlands
The Boulevard, Langford Lane, Kidlington, Oxford OX5 1GB, United Kingdom
50 Hampshire Street, 5th Floor, Cambridge, MA 02139, United States

Notices

ISBN: 978-0-12-819556-7

Publisher: Oliver Walter
Acquisition Editor: Priscilla Braglia
Editorial Project Manager: Aleksandra Packowska
Production Project Manager: Poulouse Joseph
Cover Designer: Alan Studholme

Contents

Preface

Sustainability is recognized as one of the most critical targets to achieve in the world today. Energy systems are also an integral component of this target and anything that we do directly affects it. The current global social norms rely heavily on energy systems to provide various useful energy commodities necessary for living. Such commodities include electricity, heating, cooling, domestic hot water, fresh water, and transportation fuel. These energy systems have been evolving from conventional fossil fuel and nuclear-based energy systems to more environmentally benign energy systems, such as renewable energy systems. Hybridization allows for more than one source of energy to combine allowing flexibility and creativity in the design of new energy systems based on the need and the available resources.

This book is designed to benefit students, researchers, scientists, and practicing engineers for better understanding of sustainability and assessment of energy systems from a sustainability perspective, using various thermodynamic fundamentals. This book follows the 3S concept (source—system—service), which was originally developed by the lead author, Ibrahim Dincer, where energy sources are discussed thoroughly, followed by a detailed investigation of a variety of energy systems and energy services that basically mean useful commodities. Fundamental aspects of energy, environment, and sustainability are discussed in detail in the first chapter. Sustainability modeling features various aspects including economic, social, and environmental aspects in addition to other domains such as energy, exergy, technology, education, and sizing. Moreover, although cities are in need of optimizing their energy infrastructure and resources, community energy systems is a chapter that is dedicated to discussing systems pertaining to communities as a whole. This book provides models, descriptions, analyses, and assessments of various systems and case studies.

This book is composed of eight chapters, starting with introductory information on energy forms, history, and essential thermodynamic concepts. This chapter also introduces environmental impact, climate change and global warming, as well as the relationship between energy and the environment followed by an introduction about sustainability. Chapter 2 discusses all energy sources, including primary, secondary, and converted sources where it is organized by discussing fossil fuels first, followed by nuclear energy and finally renewables. Energy systems are investigated in detail in Chapter 3, where the energy systems, particularly power generating systems, including all types of power plants are presented. Various models are described and illustrated in detail in this chapter. Other types of energy systems include heating systems and all their types such as geothermal, biomass, and heat pumps. Refrigeration systems are also explored followed by refineries. Chapter 4 dwells on energy services, particularly aiming to present useful outputs of the systems under services. Chapter 5 discusses community energy systems where combined heat and power solutions are considered, in addition to microgrids, hybrid energy models, district heating, cooling, and thermal energy storage. Cogeneration, trigeneration, and multigeneration systems are also included in this chapter. Chapter 6 focuses on sustainability models, covering eight different aspects of sustainability and the sustainability methodology used for assessment in this book. Each aspect has a number of indicators associated with them. Chapter 7 includes a number of case studies including micro and macro energy systems for small residential dwellings and community energy needs. Finally, Chapter 8 is a type of wrap-up chapter for the book, focusing on future directions and main findings of this particular book.

We hope that this book provides eye-opening type materials for the energy community and serves as a useful source for research, innovation, and technology development in the field of energy sustainability.

Ibrahim Dincer
Azzam Abu-Rayash
July 2019

CHAPTER

Fundamental aspects of energy, environment, and sustainability

1

1.1 Introduction

Energy plays a pivotal role in the development and prosperity of nations. In fact, the industrial revolution, followed by the oil explorations combined, makes up our current digital civilization. Furthermore, aside from power, energy influences our lives on a daily basis. For example, electricity infrastructure, the transportation, and industry sectors all depend on energy. In fact, Holden et al. (1997) published a book discussing the political economy of South Africa through its transition from minerals-energy complex to industrialization. In this book, energy is a driving factor in the economy for South Africa and the rest of the world, which consequently becomes a major factor in political dynamics. Moreover, Georgescu-Roegen (2018) dwells in detail to highlight the limitation of natural resources and their impact on global economy. In this chapter, the author analyzes energy options and discusses in detail the degree of influence each aspect has on the global economy. On the other hand, Gomez-Exposito et al. (2018) focused on the electric aspect of energy systems by providing a deep and a comprehensive understanding into modern electric energy systems. Topics of research in this field include renewable penetration, smart grids, and active consumption. Furthermore, electrical aspects of research include harmonic analysis, state estimation, optimal generation scheduling, and electromagnetic transients. Babu et al. (2013) have summarized the hydrate-based gas separation process for carbon dioxide precombustion capture. Superhydrophobic surfaces are also a recent topic of research for various energy-related applications including heat exchangers, ice slurry generation, photovoltaic cell, electric power line, and airplanes (Zhang and Lv, 2015). These devices benefit from the freezing delay and the avoidance of ice accumulation on surfaces to maintain operational function. In addition, latest research also revolves around the use of hydrogen as an energy carrier or source for various systems. Nastasi and Lo Basso (2016) investigated the use of hydrogen as a link between heat and electricity in the transition toward future smart energy systems. The duality in the use of hydrogen as both a fuel for combustion and a chemical for energy storage or chemical conversion along with its abundance gives it a unique feature above other energy options. Additionally, energy storage is also another hot topic for research. Luo et al. (2015)

Energy Sustainability. https://doi.org/10.1016/B978-0-12-819556-7.00001-2

investigated the current development in electrical energy storage technologies and their application potential in power system operation. This features the dynamic changes of the grid system along with the mixed energy sources in modern electric grids as well as the reduction in natural resource and the exponentially increasing population of the world. Moreover, energy storage systems for wind power integration support are investigated by Zhao et al. (2015). Furthermore, smart energy systems have been analyzed for 100% renewable energy and transport solutions by Mathiesen et al. (2015). They identified least cost solutions of the integration of fluctuating renewable energy sources. In addition to renewable energy, utilization of various fossil fuel by-products such as carbon dioxide and natural gas hydrates are being researched. Chong et al. (2016) reviewed the natural gas hydrates as an energy resource. Moreover, dark fermentative biohydrogen production from organic biomass including agricultural residues, agro-industrial wastes, and organic municipal waste has been investigated by Ghimire et al. (2015). In fact, further research and development to this technology include improving the biohydrogen yield by optimizing substrate utilization, microbial community enrichment, and bioreactor operational parameters such as pH, temperature, and H_2 partial pressure. As for research around renewable energy, a technical and an economic review of renewable power-to-gas process chain is investigated by Götz et al. (2016) and is thought to play a significant role in the future energy systems. In this process, renewable electric energy can be transformed into storable methane via electrolysis and subsequent methanation. Furthermore, the potential of lithium-ion batteries in renewable energy is further analyzed by Diouf and Pode (2015) as a major energy storage medium for off-grid applications. Moreover, the integration of renewable energy systems into the future power systems is researched in detail by Weitemeyer et al. (2015). A modeling approach to investigate the influence of storage size and efficiency on the pathway toward a 100% RES scenario is presented after using a long-term solar and wind energy power production data series. Overall, the main objectives behind latest energy research are to develop environmentally benign energy solutions as well as improve energy storage options for more sustainable and reliable energy supply from renewables. Furthermore, the environmental, social, and economic aspects of energy drive the sustainable development of all energy systems. Moreover, unprecedented records of high global temperatures and the universal climate change have been a major trigger to becoming more environmentally conscious, which eventually drives energy research in this direction.

1.2 Energy

Energy is an important constant of the universe. Energy is the ability to do work, whereas work is the active displacement of an object by applying force. Energy seems near tangible to us, as it is present in daily activities. This is because energy is not a substance or an element, but rather a quantity, derived from a mathematical relationship with other more fundamental quantities. Therefore, because energy is a conserved quantity, energy cannot be created or destroyed, rather can be converted in form according to the law of conservation of energy. The SI unit used to calculate energy is joule, which is the energy transferred to an object by exerting a force of 1 N against it while moving it a distance of 1 m. On the earth, most of like is powered by a central source of energy, the sun. Radiant energy from the sun is emitted into space after the sun is heated to high temperatures due to the conversion of nuclear binding energy. Moreover, energy comes in various forms such as kinetic, potential, elastic, chemical, gravitational, electric, magnetic, radiant, and thermal energy. Consequently, energy has numerous applications on every segment of life around us. Therefore, energy is very valuable as it affects us daily. Table 1.1 summarizes the main introductory aspects of energy.

1.2.1 Energy forms

Energy can be classified into two main categories: kinetic and potential energy. Kinetic energy refers to the energy that an object possesses due to its motion. Maintaining the acceleration, the objects keep their kinetic energy. On the other hand, potential energy reflects the potential of an object to have motion and it is generally a function of its position relative to the surrounding field. The interaction between kinetic and potential energies results in many types of energy. Fig. 1.1 illustrates various types of energy that result from the combination of kinetic and potential energies.

Therefore, energy can manifest itself in many forms. In fact, energy can be converted from one form to another depending on the need and available resources. To elaborate further on these types of energy, Table 1.2 presents the different types of energy along with a short descript of each and a common application for each type.

1.2.2 Energy history

In the 17th century, Gottfried Leibniz defined the mass of the object and its velocity squared as *vis viva*, or living force. Later in 1807, Thomas Young used the term "energy" instead of *vis viva,* which then was described as kinetic energy. Later, William Rankine devised the term potential energy. Shortly after, the law of energy conservation was postulated in the early 19th century. In 1845, James Joule discovered the link between mechanical work and heat generation. All of these developments have

Table 1.1 Highlighted summary of basic energy properties.

Term	Unit
SI unit	Joule ($J = kg\ m^2/s^2$)
Other units	Calorie, kWh, kcal, BTU

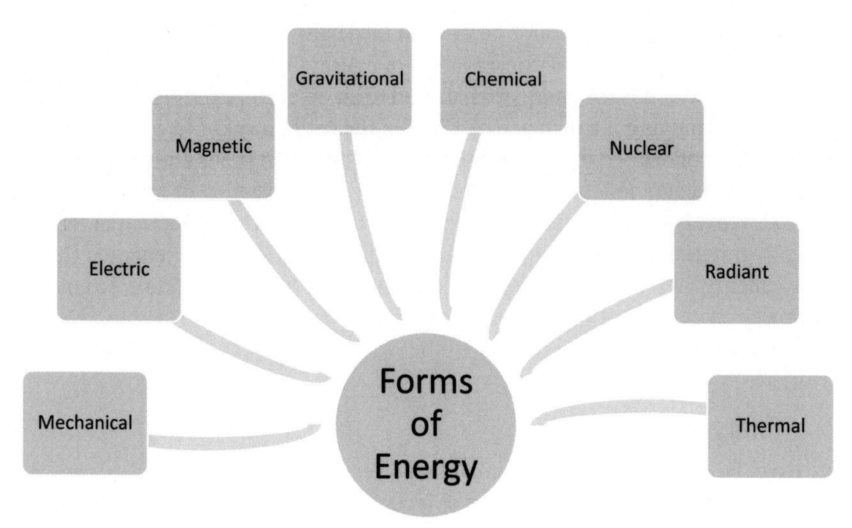

FIGURE 1.1

Forms of energy emerging from the combination of kinetic and potential energies.

Table 1.2 Descriptions of various types of energy and example application for each type.

Type	Description	Application
Mechanical	Energy acquired by objects, where work is done upon them.	Using a hammer to apply force on a nail.
Electric	Kinetic energy of moving electrons.	Electric power utility.
Magnetic	Energy by magnet in a magnetic field.	Operation of transformers and inductors.
Gravitational	Potential energy of an object due to its high position.	Object falling from top of a building down to the ground.
Chemical	Energy stored between atomic bonds.	Chemical reactions such as burning sugar in human body.
Nuclear	Nuclear reactions releasing nuclear energy to generate heat.	Nuclear power plants.
Radiant	Energy of gravitational and radiomagnetic radiation.	Solar energy.
Thermal	Internal heat within a system responsible for its temperature.	Warming up a kettle.

led to the theory of conservation of energy, which was later formalized by Lord Kelvin as the field of thermodynamics. This emerging field of thermodynamics aided the investigations of chemical processes as well as led mathematical formulations of the concept of entropy.

Energy played an integral and significant role in the development of civilizations as it influenced the social, economic, and geopolitical aspects of societies across the globe. The destruction of modern or ancient societies was due to a number of factors that directly or indirectly connect to the shortfall of energy resources. Collapse was accelerated by wars that emerged due to competition over scarce energy resources. Large energy absorption can lead to the destruction of civilizations according to history. Imagine what could happen to our modern society that is founded on exaggerated energy consumption?

The basic needs for humans energy-wise remains unchanged: heat, light, manufacturing, and transportation. Humanity has passed through numerous stages of energy development. Each human society had relied on energy in its distinct form. However, the past two centuries have been unprecedented in human history in terms of energy development and transformation. The intensity and advancement of energy today is a major milestone in human history. This great acceleration on environmental impact and economic changes is historic. Modern lifestyle is radically different from the lifestyle of our ancestors. Our societies have been extremely dependent on energy. For example, lighting at night is considered a necessity and a service that is readily available in the 21st century. In the past, people struggled with lighting indoor amenities with candles or dried strips of vegetables dipped into animal fat and thus producing a filthy smell. Once the sun set, they no longer had night illumination or street lights all over their cities. Heating was only provided in extreme cases of necessity. Even well-to-do figures struggled with ink freezing in the inkpot. Now, manufacturing services run at all times. Central air conditioning and temperature control has become a norm in our society, thus cooling indoor space in the hot summer days and heating them in the harsh winters. These services can be seen as the same services provided in the past (i.e., heating, lighting, etc.) to run the daily affairs. However, the shock is that none of these services comes from the same sources as they did in the 19th century. Not even one. Such a paradigm shift has never been witnessed since humans learned to harness fire. Heat, light, and motion not only provide us with necessary services but also provide a wide range of new services that have become available to us such as pictures that come from screen or music from speakers. As a result of such remarkable transformation, our command over resources has got much stronger as individual human beings and societies have expanded. Furthermore, the degree of choices available to us has immensely increased.

About 2 million years ago, humans learned to manufacture tools for hunting. Around 500,000 years ago or earlier, humans discovered the use of fire. Fire was used to create and to destroy. It provided light and warmth and also was used as a weapon to kill. With this new resource, humans were able to shape their environments by selectively creating or destroying what was necessary for their survival. Late on, humans learned to use the fire for craft (i.e., melting metals or hardening clary), which enabled them to trigger more environmental changes. About 18,000 years ago, the animal power was a major source of power. Domestication of animals enabled humans to capture a new substantial energy source. Sheep, goats, and cows provided not only a reliable food source but also a predictable mobile source of energy for nomad populations. Around 10,000 years ago, some nomad populations began to settle near rivers and fertile land to develop more reliable and consistent sources of energy. Permanent settlements and the population growth urged human societies to rely on the resources around them from trees, soil, water, and animals to satisfy their needs. Around 8000 years ago, the use of animals to pull carts began, which caused agriculture to flourish. Stationary tasks such as milling of grain, pumping of water, and other mechanical conversions of energy only emerged in the 19th century. In 2000 BC, early missionaries to China have reported that coal was already being used for heating and cooking, making the Chinese to be the first to use coal for energy use. The report also mentions that coal has been utilized for more than 4000 years. In another report, Marco Polo highlights the widespread use of coal in China in the 13th century. Indeed, coal might have been used by mammoth hunters in Eastern Europe. In medieval Europe, the existence of coal was also recognized, yet ignored due to the soot and smoke. Wood was favored over coal until the 13th century. The Greeks also recognized coal from a geologic curious point of view. Aristotle mentioned coal, and the context implies that he was referring to it as a mineral of earth not as an energy source. During the Bronze Age, coal was used in southern Wales. The Romans used coal in large quantities in multiple locations. After the Romans have left, the use of coal stopped until the second millennium. Lignite and peat are geologic precursors to coal, and they were used in northern and Western Europe in the first century AD. Indications point at the Netherlands, where peat was used as a fuel. Romans burned coal near St. Etienne, which later became a major French mining center. Coal mining in India traces back to the 18th century. However, names and signs in the Bengal-Bihar region indicate that coal may have been used in these areas in ancient times. The first practical use of natural gas is also traced back to the Chinese in 200 BC, where they used to make salt from brine in gas-fired evaporators, boring shallow wells, and conveying the gas to the evaporators through bamboo pipes. In the same era of 200 BC, Europeans harness water energy to power mills. The invention of this vertical waterwheel powered mills, which refreshed various industries. It also decreased the dependence on human and animal muscle for the production of power. Furthermore, sites with decent waterpower potential have become more favorable, and communities started to be established around these places, causing economic, industrial, and social growth. In the first century, the Chinese have refined petroleum to use it as an energy source. Sheng Kuo (1031−1095) have documented that there was a lot of oil in the subsurface and that it was inexhaustible. They applied petroleum for lamps, as lubricants in medicine, and other uses. In the 10th century, windmills were built in Persia to grind grain and pump water. This technology spread to China and the Middle East, where farmers used them to irrigate crops, pump water, and crush sugarcane.

The history of energy continues to our current modern societies, where renewable energy is on the rise and fossil fuels are influencing the environment significantly. Moreover, energy consumption per capita varies from country to another. Fig. 1.2 demonstrates the annual average energy consumption per capita for the 20 largest energy-consuming countries in the world along with the global average.

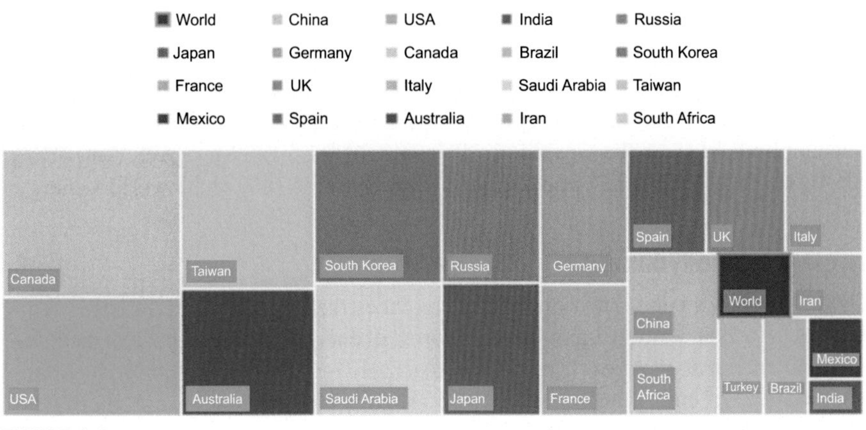

FIGURE 1.2

Annual average energy consumption per capita in 2016 (kWh/per/yr).

(Data from: IESO, 2016).

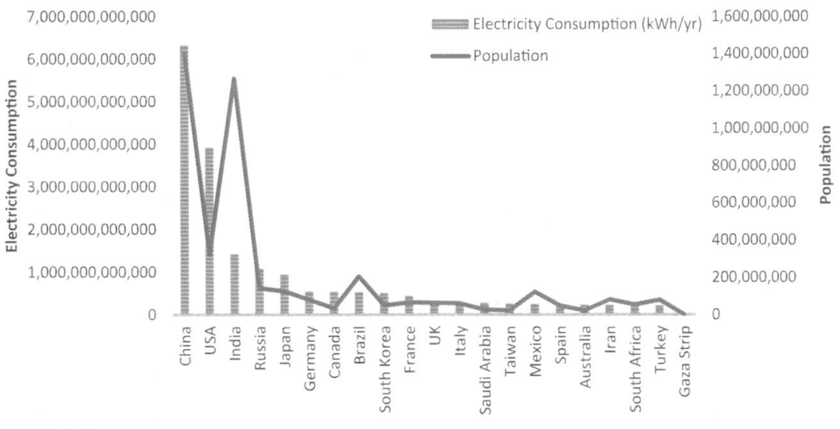

FIGURE 1.3

Electricity consumption versus population for various countries across the world for the year 2014.

Data from Ontario Energy Report Q1 2016. Independent Electricity Service Operator, Toronto, ON, pp. 1–16, Rep.

It is noticeable that energy consumption is not equally distributed. In fact, the United States, China, India, and Russia are the most consuming countries. Furthermore, the 20 largest energy consumers account for 80% of total primary energy consumption, with the two largest consumers (the United States and China) accounting for 40% of the total global consumption. Furthermore, Fig. 1.3 demonstrates the relationship between population and electricity consumption in the top 20 countries worldwide as well as the lowest electricity consuming strip on the planet, the Gaza strip.

Fig. 1.3 is very concerning as it demonstrates inequality in the global energy consumption and distribution. The poorest 10% account for 0.5% of total global energy consumption, whereas the wealthiest 10% account for 59%. Moreover, Americans constitute 5% of the world's population and yet consume 24% of the world's energy.

1.2.3 Thermodynamics

Thermodynamics is a branch of physical sciences that specializes in heat and temperature and their relationship with other forms of energy such as electrical, mechanical, and chemical energies. Dincer and Acar(2018) proposes that thermodynamics is the science of energy and exergy. There are four fundamental laws of thermodynamic that define the physical quantities of a thermodynamic system. Moreover, thermodynamics feature four branches including classical, statistical, chemical, and equilibrium thermodynamics. The main laws of thermodynamics are as follows:

a. *Zeroth law of thermodynamics*: If two systems are each in thermal equilibrium with a third, then they are also in thermal equilibrium with each other.
b. *First law of thermodynamics*: The internal energy is constant for isolated systems. Energy is neither destroyed nor created. It is always conserved.
c. *Second law of thermodynamics*: Heat cannot flow spontaneously from a colder location to a hotter location. Exergy cannot be conserved. It can only be minimized if the measures are taken properly.
d. *Third law of thermodynamics*: As a system approaches absolute zero, all processes cease and the system's entropy approaches a minimum value.

The first law of thermodynamics is the main law highlighting conservation of energy. It implies that although energy can change its form, it can neither be created nor be destroyed. Therefore, this law is closely associated with the energy aspect of any system. However, this law does not give information on the direction of which the processes spontaneously flow. This aspect refers to the reversibility of thermodynamics. This is covered by the second law of thermodynamics, which embodies the ability to assess the energy qualitatively, characterize the availability of required energy, and specify reversible reactions. Indeed, the second law is closely associated with the exergy aspect of any system.

Therefore, the first and second laws of thermodynamics are the constitutional laws that drive the discipline of thermodynamics, and whereas the first law is a measure of quantity only, the second law is a measure of both quantity and quality. Exergy and the translation of the second law of thermodynamics is necessary, as in reality we fail to identify wastes or use fuels effectively. In addition, the second law of thermodynamics and consequently exergy aim to introduce better efficiency, cost-effectiveness, design, analysis, and implementation as well as better strategies and policies. Moreover, exergy positively influences the environment and in turn sustainability and energy security.

In addition, the concept of exergy is critical when analyzing sustainability aspects of energy systems. The impact of exergization not only is limited to environmental friendliness but expands to economic performance and more effective and efficient energy systems and sources. For example, Fig. 1.3 shows a person with a huge battery, to simply power his phone. Employing exergy as a tool toward energy sustainability allows for using green energy and cleaner technologies as well as conserve energy. This conflicting dilemma is an example proving that importance of exergy. Large-scale implications of this figure can be postulated when using various electricity generation plants for urban use, for example, and which energy mix is most suitable and practical for the services and commodities needed. Fig. 1.4 also demonstrates a thought-provoking question about the concept of exergy and its role in energy sustainability. This nicely illustrates the concept of exergy by meaning that exergy as a tool help make energy systems more environmentally benign. It was presented in a Parliament Research Day Event in Toronto, Ontario to display Dr. Dincer's research.

Moreover, in thermodynamics, properties combine to create a state point, which can consequently be combined to form a process. More than two processes can be designed to formulate a cycle. Fig. 1.5 illustrates the six-step approach in thermodynamics (Dincer and Acar, 2018).

Thermodynamic properties include temperature, pressure, volume, specific internal energy, and specific entropy. These properties are used to identify any selected state point. The state point is an identified point within the cycle with distinct properties. For example, points are distinct before and after a given heat exchanger as the temperature and, consequently, the enthalpy are different before and after a fluid passes through a heat exchanger. Furthermore, a process is defined by at least two state points. The actual change from one state point to the next is considered a process. In fact, two more processes in a closed-loop sequence make up a cycle. Balance

FIGURE 1.4

Significance of exergy using creative thought-provoking illustrations (Dincer, 2016)

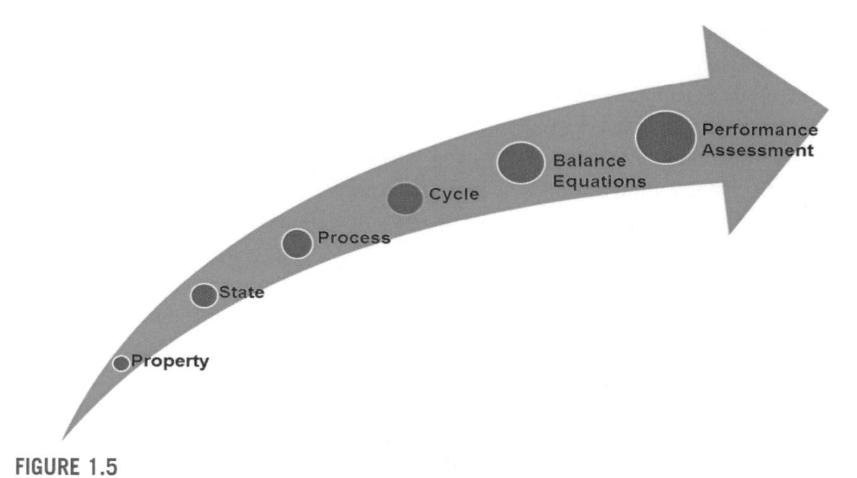

FIGURE 1.5

Approach to thermodynamics using the six-step approach.

equations are then carried out to configure the mass balance equations, energy balance equations, exergy balance equations, and entropy balance equations, taking into account the different outputs and inputs of the overall system or cycle. After configuring the balance equations, thermodynamic performance assessments can be conducted by analyzing the energetic and exergetic efficiencies of the system as well as exergy destruction and losses and other thermodynamic parameters.

1.3 Environment

Environment refers to the physical and natural environments that encompass all living and nonliving organisms. This term can also refer to the earth, or some parts of it. The concept of environment also encompasses the interaction between all living species, climate, natural resources, weather, and human activity, which affect human survival and economic prosperity. The reason why the environment is significantly highlighted in the book is the current phenomenon, where earth and environmental processes are heavily impacted by human activity. In fact, hundreds of years ago, human activity had little or insignificant impact on earth, and human needs and lifestyles were in harmony with natural processes. However, modern societies and current lifestyles have evolved very rapidly to the point where environmental processes and earth itself are changing because of various human activities.

1.3.1 Environmental impact

Pollution is the introduction of contaminants into the natural environment, which causes adverse change. Pollution can be of many forms such as chemical substances or energy, such as noise, heat, or light. Toxicants are any toxic substances that may be man-made, manufactured, or biologically produced. Pollution and toxicants can be observed in the air, soil, water, or food. Although policies and bills are being passed to combat pollution, the problem remains at present. Fig. 1.6 shows the different types of pollution. Pollutants are not only harmful to the environment, but some may have health and social effects as well. In fact, adverse air quality can kill many organisms including humans. Ozone depletion can also cause various respiratory complications, cardiovascular illnesses, and throat inflammation. Noise pollution can also cause hear loss, high blood pressure, and stress. Moreover, sulfur dioxide and nitrogen oxides introduced the phenomenon of acid rain, which lowers the soil pH, making it infertile and unsuitable for crops. Carbon dioxide and greenhouse gas (GHG) emissions also have significant impacts on ocean acidification and global warming, which adversely affects many ecosystems on earth. Therefore, pollution and contaminants are deadly when it comes to the social and health aspects.

In fact, the total emissions worldwide for the year 2012 have totaled 51,840 Mt CO_2-eq. The industrial sector accounts for almost one-third of the total emissions. Furthermore, coal is the energy source with the highest emissions. It is primarily burned for electricity generation in the steel and cement industries. Direct emissions include methane emitted by cows and other livestock, logging or cutting down trees, or organic matter in landfills. Natural gas is normally used as an energy source for cooking, heating, and electricity generation. Furthermore, oil emissions result from combustion as transport fuels in cars, trucks, and airplanes. Fig. 1.7 shows the global CO_2 emissions by source and sector for the year 2013.

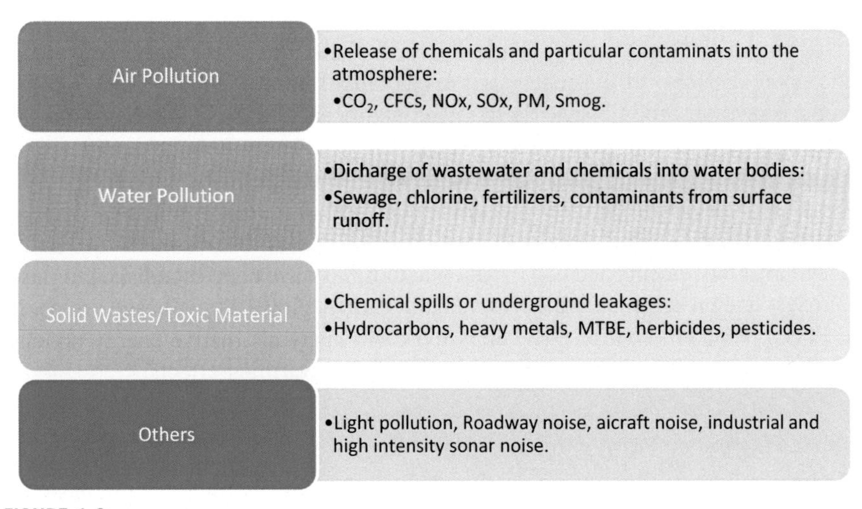

FIGURE 1.6

Major types of pollution and contaminations along with common pollutants.

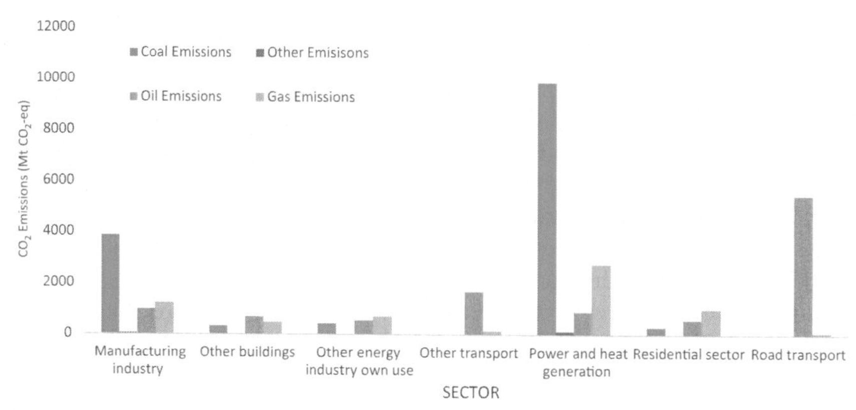

FIGURE 1.7

Global CO_2 emissions by source and sector for the year 2013.

Data from IESO, 2016.

1.3.2 Climate change and global warming

Global warming and climate change is a global trend, stemming from a century-scale rise in temperature of the earth's climate. The increase of temperature is primarily due to energetic activities such as the GHG emissions. The greenhouse effect occurs after the burning of fossil fuels and results in massive amount of heat and energy in the atmosphere. The main GHG pollutant is carbon dioxide, and often methods of measuring the global warming potential confine to evaluation of carbon dioxide. However, other major pollutants contributing to global warming and climate change include, but not limited to methane, nitrogen oxides (NOx) and sulfur oxides (SOx). Although some thermal radiation is emitted into space, the bigger part of this residual heat is absorbed by the earth. Furthermore, the weakening of the atmosphere disables it from reflecting the incoming solar radiation. Rather, this radiation is also absorbed by the earth's land and ocean surfaces, causing unprecedented warming of the earth. These events are triggers to critical ramifications such as arctic sea ice melting, rise of sea levels, glacier retreats, and many atmospheric anomalies. To mitigate the impacts of global warming, there have been numerous efforts socially, politically, and scientifically to resolve this issue. Environmentalist movements and conservation authorities have been established to raise awareness and preserve natural environments. Various global summits and protocols have been passed to control GHG emissions, and many alternative energy options have been proposed and deployed to mitigate global warming effects.

1.3.3 Energy and the environment

The combustion of fossil fuels such as coal and oil are the primary produces of GHGs. Therefore, energy consumption is the main cause of climate change. In fact, before the discovery of fossil fuels and before their deployment in industrial and commercial uses, human activity and lifestyle had little or insignificant impact on earth and the natural processes. However, the industrial revolution and later the discovery of oil have revolutionized the human civilization, as such energy sources play a vital role in economic growth, energy security, and political strength. Furthermore, advancements in energy research such as the development of alternative fuels and renewable energy systems were primarily triggered by the relationship between energy and the environment. In fact, the concern of global warming and climate change urged the development of energy through alternative means and through renewable sources. Furthermore, the relationship between energy and the environment is very closely linked, as all energy is either derived or captured from the environment around us. Once energy has been used, it also returns to the environment, either as a harmless by-product or as a harmful waste. In summary, energy and the environment have been always closely linked and will remain as such in the future.

1.4 Sustainability

Sustainability has become a major highlight in modern civilization. In fact, a crucial phenomenon, sustainability, which is always present in political debates, educational programs, social trends, and scientific advancements, has become multidisciplinary. However, the concept of sustainability has always been present throughout human civilizations. Indeed, humans always planned, and their concern for resource availability was on the top of their priority list. This was the reason for humans to shift from hunting into farming. *Nachhaltigkeit,* a German coin that referred to "sustained yield," was the original term for sustainability, which was found in a forestry book in 1713. Sustainability does not have a standard definition. Scientists defined sustainability in a variety of ways, depending on the context and the scientific field of use. In the Oxford dictionary, sustainability is defined as the avoidance of the depletion of natural resources to maintain an ecological balance. Encyclopedia Britannica defines sustainability as the long-term viability of a community, set of social institutions, or societal practice. Table 1.3 provides a list of definitions that are used for sustainability. Sustainability and sustainable development can be used interchangeably, as the latter could also mean a continuous or sustained development. Development indeed is a qualitative improvement to a system, which is distinct from growth. Growth denotes to quantitative increase in physical scale.

The concept of sustainability has closely been associated with energy consumption. The early humans started to use the fire for specific foods, which may have altered the natural composition of the planet and animal species (Scholes, 2003). Civilizations then transformed from hunting to more sustainable societies by introducing agriculture. In fact, agrarian communities depended largely on their environment (Clarke and Scruton, 1977). The longevity of societies and an important factor that determined its flourishment or destruction was sustainable development. Energy is a critical element, which effects the interaction between nature and societies. In the past, increases in energy demands were associated with economic and

Table 1.3 Selected definitions of sustainability, sustainable development, or sustainability sciences.

Source	Definition
(Brundtland Commission of the United Nations, 1987)	Sustainable development is development that meets the needs of the present without compromising the ability of future generations to meet their own needs.
Pearce and Markandya (1989)	Sustainable development involves devising a social and economic system, which ensures that these goals are sustained, i.e., that real incomes rise; that educational standards increase that the health of the nation improves; that the general quality of life is advanced.
Harwood (1990)	Sustainable agriculture is a system that can evolve indefinitely toward greater human utility, greater efficiency of resource use, and a balance with the environment, which is favorable to humans and most other species.
Morelli (2011)	Meeting the resource and services needs of current and future generations without compromising the health of the ecosystems that provide them. In specific, sustainability is a condition of balance, resilience, and interconnectedness that allows human society to satisfy its needs while neither exceeding the capacity of its supporting ecosystems to continue to regenerate the services necessary to meet those needs nor by our actions diminishing biological diversity.
(Forum for the future, 2008)	Sustainable development is a dynamic process that enables people to realize their potential and improve their quality of life in ways, which simultaneously protect and enhance the earth's life support systems

technological advancements. However, currently, the rise in energy consumption may be detrimental social, environmental, and even economic effects. These effects could include local and global health impacts. The development of civilizations and the introduction of various energy sources evolved the concept of sustainability. Although early civilizations utilized limited amounts of energy, industrial societies rely on abundant energy sources. Transportation, heating, and electricity compose the main needs of humans. However, with the industrial revolution and the technological advancements, novel energy resources have shaped the modern human civilization. Coal, oil, natural gas, and other conventional energy sources have caused an exponential increase in human consumption of resources.

This in turn triggered environmentalism and the introduction of new fields such as ecology and environmental sciences. The unprecedented increase in energy demand, population, and economy led the world to realize the impacts of energy use on the economy, environment, and socially. Energy conservation, sustainable energy options, and renewable energy sources were therefore explored further. Indeed, this concern has been widely addressed globally. The united sustainability in its modern context refers to a relatively complex topic that is multidisciplinary. The meaning could vary depending on the context and the field. Indicators of development toward sustainability provide meaningful data that could characterize systems' sustainability level. Sustainability assessment relies on a number of indicators. These indicators could differ from one study to another. Concerned about economy, energy, society, and environment, governments have been introducing regulations and local/regional bylaws that aim to enhance sustainability measures. However, these efforts often include nonrigorous assessment methods that are mainly qualitative in nature.

Sustainability is a complex and interdisciplinary concept, which relates to each of the presented domains in Fig. 1.8. To assess the sustainable development of energy systems, resources, energy, and the economy must be taken into detailed consideration. Furthermore, the environmental footprint, social impact, cultural paradigms surrounding these programs as well as public policy and political aspects need to be studied thoroughly to comprehensively and objectively understand sustainability.

Moreover, some of these concepts are interdependent. For example, the economic domain could influence the social and public policy domains. Overall, the road toward an objective understanding and assessment of sustainability springs

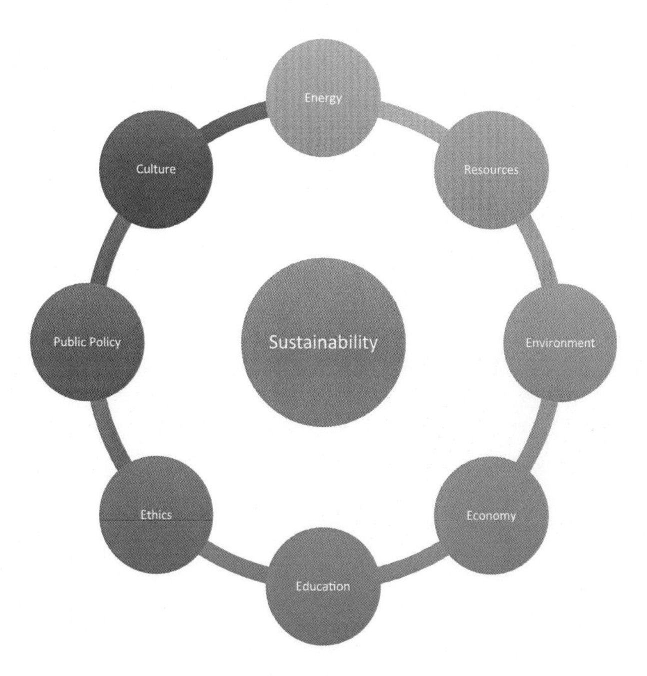

FIGURE 1.8

The backbone of sustainable development and the major domains that contribute to the understanding of the sustainability concept.

from sound and deep analysis of all factors and elements that contribute to this concept whether directly or indirectly. The subjectivity surrounding sustainability make poses as a disadvantage to its essence as well as an opportunity for creativity and innovation. In fact, sustainability can be applied to various disciplines and concepts from energy systems to corporate human resources. As for energy sustainability, various technical parameters must be clearly defined to provide meaningful results along with the consideration of other important aspects such as public policy, culture, education, and ethics. The use of sustainability and its applications in various disciplines can be better understood through the illustration in Fig. 1.9.

Moreover, energy sustainability can be illustrated by using the 3S approach developed by Dincer and Acar (2017). The 3S approach presents sustainability by breaking the various aspects of energy including source, system, and service and furthermore introducing emphasizing storage option in between each aspect. The 3S approach as illustrated in Fig. 1.10 stresses that the energy source must be clean, abundant, cheap, and available to achieve overall energy sustainability. Once energy is harvested, its effective storage solutions are proposed before moving to the next aspect of energy, which is the utilization of energy using various energy systems.

Moreover, to maintain energy sustainability, systems need to be highly efficient, effective, and reliable as well as be able to recover losses and wastes. Once more, storage solutions are important before dispatching the energy directly to the service.

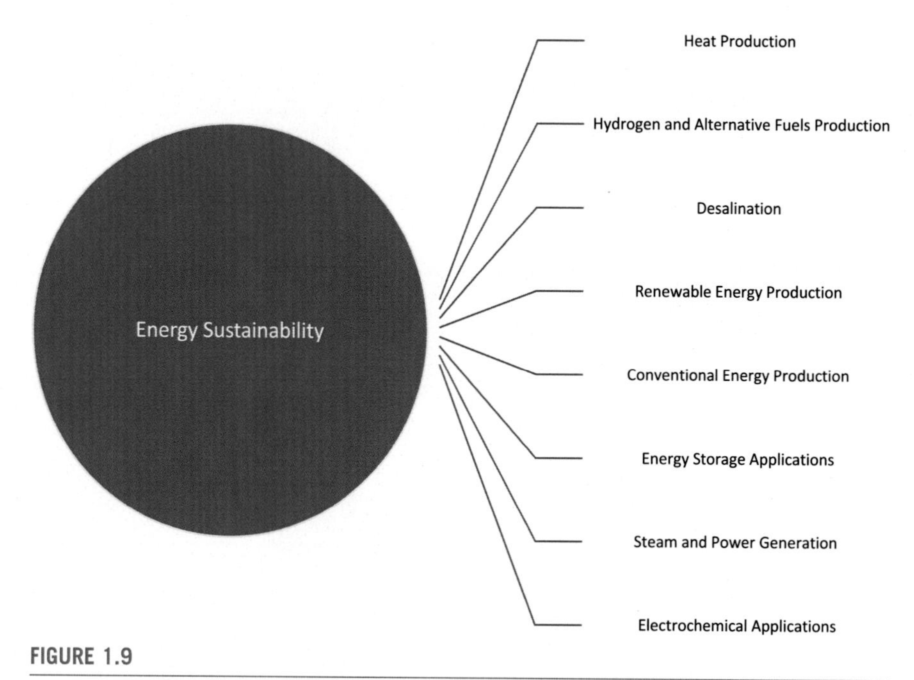

FIGURE 1.9

Illustration that highlights the use of sustainability in various energy production disciplines and applications.

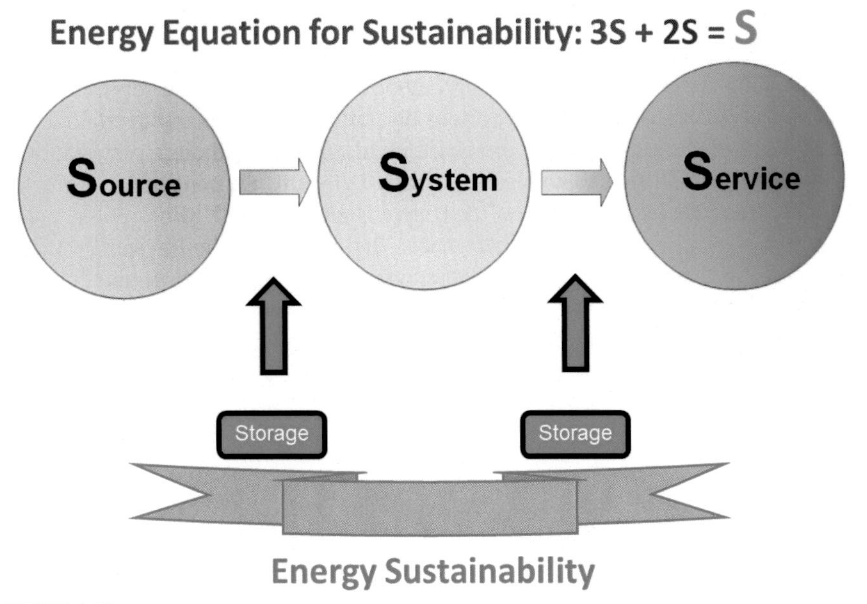

FIGURE 1.10

The 3S concept that was originally introduced by Dincer (2015) to achieve energy sustainability.

Finally, sustainability is achieved by providing clean, practical, and efficient services. The 3S rule is critical when analyzing any energy system. The flow of energy from the source to the system and finally to the service and the integration of storage solutions is both logical and practical. Energy sustainability can be achieved at each step in this process. For instance, sustainability at the energy source level can be reflected by utilizing clean, cheap, and available energy resources versus the opposite. On the other hand, sustainability at the system level is achieved by wasteless and efficient energy systems. Multigenerational energy systems always yield in higher efficiencies and less energetic and exergetic losses. Lastly, the service sustainability is by providing a reliable and clean energy service. This energy equation for sustainability therefore is denoted by five small (s) that represent source, system, service, and the intermediate storages, all yielding to the ultimate result of a large (S) that denotes Sustainability or energy sustainability.

1.5 Closing remarks

The fundamental aspects of energy, environment, and sustainability have been outlined in this chapter. As indicated, environmental impacts of energy solutions have driven the research around energy sources, storage, and energy utilization. Furthermore, details around environmental pollution, climate change, and the concept of sustainable development have been discussed to give an introductory foundation for this book. The energy equation for sustainability is also introduced to give a better understanding about energy sustainability at the source, system, and service levels. The application of sustainability in various disciplines and the complex multidisciplinary feature of this concept are presented in this chapter. Although sustainability has numerous definitions and none is standardized globally, the book adopts the definition by Brundtland Commission of the United Nations in 1987, which defines sustainability as the development that meets the needs of the present without compromising the ability of future generations to meet their own needs. Furthermore, the concept of exergy is introduced as a critical tool in determining energy sustainability. In closing, the six-step approach to thermodynamic assessment of energy systems is introduced in this chapter to give a better understanding about thermodynamic analyses.

CHAPTER

Energy sources

2

2.1 Introduction

Although energy is extracted from the environment through various means, some sources are more prevalent than other sources. In fact, some sources were most widespread such as coal until the discovery of oil. Currently, coal is seldom used as an energy source in the presence of oil and natural gas because of its heavy environmental footprint. Furthermore, renewable energy sources are also being developed and are widespread to offset the polluting and depleting fossil fuels. Fig. 2.1 shows a list of energy sources that have been used to meet the demands of various commodities.

Energy resources can be classified in many ways. One of the ways is the classification of energy resources into primary and secondary sources. Primary sources are those that can be suitable for end use without conversion into another form. For example, solar, wind, coal, oil, natural gas, and uranium are all examples of primary energy resources. On the other hand, secondary resources are those such as hydrogen, synthetic fuels, or hydrogen. Another classification of energy resources is the time required to regenerate an energy source. For example, renewable energy sources are ongoing and not depleted by human activity. On the other hand, nonrenewable energy sources such as fossil fuels are those that are significantly depleted by human activity and will less likely recover their potential during human lifetime. Energy portfolios are very complex, as primary sources of energy include a diverse variety of fuels. Fig. 2.2 illustrates the high-level summary of the energy sector along with the transformation of primary sources and finally the end uses.

2.2 Fossil fuels

Fossil fuels make up the bulk of the world's primary energy sources. In fact, these fuels have been the highlight of the global stage since the industrial revolution. Simply, fossil fuel sources burn coal or hydrocarbons, which are the remains of the plant and animal decomposition. Heat generated from the burning is used directly for space heating or further processed through mechanical energy to produce electricity and other commodities. Fossil fuels account for more than 70% of the world's

Energy Sustainability. https://doi.org/10.1016/B978-0-12-819556-7.00002-4

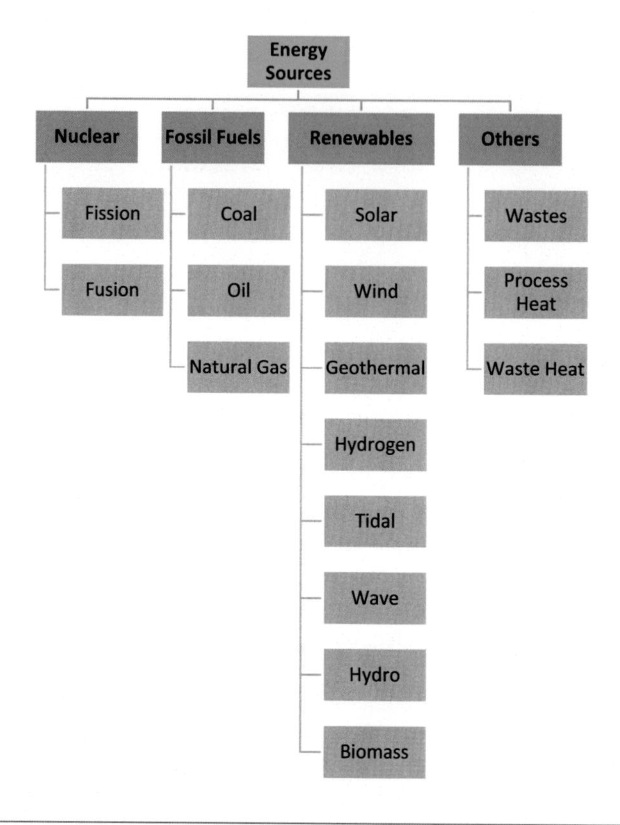

FIGURE 2.1

Classification of energy sources and list of sample primary sources.

energy in 2017 (IEA, 2018). Naturally, the commercial viability for fossil fuels is well established, as the infrastructure and technologies for these sources are mature. Furthermore, the energy dependence on imported fossil fuels imposes risks on countries that do not have such resources in abundance. These dependencies have led to monopolization, wars, and instability. Lastly, the combustion of fossil fuels results in significant greenhouse gases (GHGs) including carbon dioxide, nitrogen oxides, sulfur oxides, carbon monoxide, particulate matter, mercury, arsenic, volatile organic compounds, and heavy metals. Indeed, fossil fuels are not environmentally friendly and have been the leading cause of a global climate change. Fig. 2.3 shows historic fossil fuel production between 1981 and 2015 in the world. The reliance on oil and coal remains an environmental concern. The steady increase in fuel production reflects the increase in global energy demand and world population. As global population and energy demand increase, sustainability of the energy sector becomes critical to maintain lifestyles and essential commodities. Therefore, other energy sources must become available while more competent energy systems are upgraded regularly to ensure highest efficiency and performance.

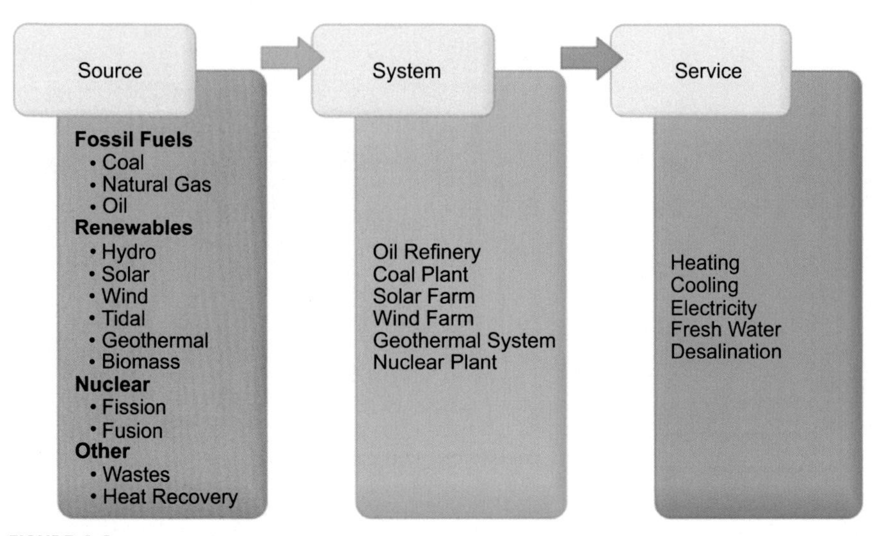

FIGURE 2.2

The flow of energy from primary sources to end uses following the 3S model.

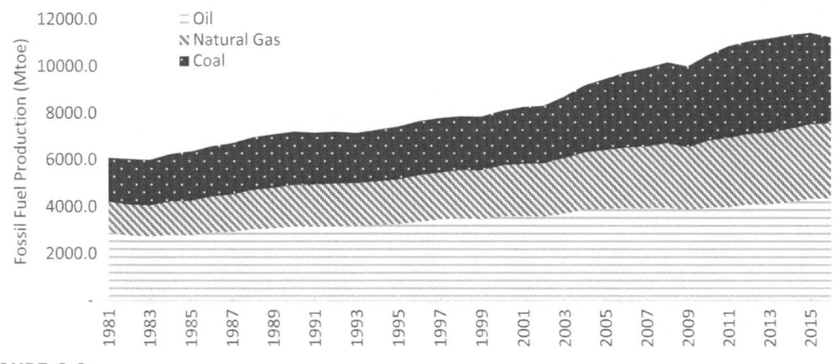

FIGURE 2.3

Historic fossil fuel production in the world between 1981 and 2015.

Data from: BP Statistical Review of World Energy. 2017. BP, London.

Although the fossil fuel—based energy production increases steadily, the global energy consumption continues to grow at a sharper pace while incorporating various sources of energy aside from fossil fuels. Fig. 2.4 shows the historic energy consumption per fuel between 1965 and 2015. Although the base load of the world's energy composition is primarily oil and coal, nuclear and renewables are becoming more widespread at a faster pace.

Furthermore, Fig. 2.4 shows that fossil fuels still make up the base loads for electricity and energy consumption. In specific, oil, natural gas, and coal combined almost make up the majority of energy sources for the world historically. This is

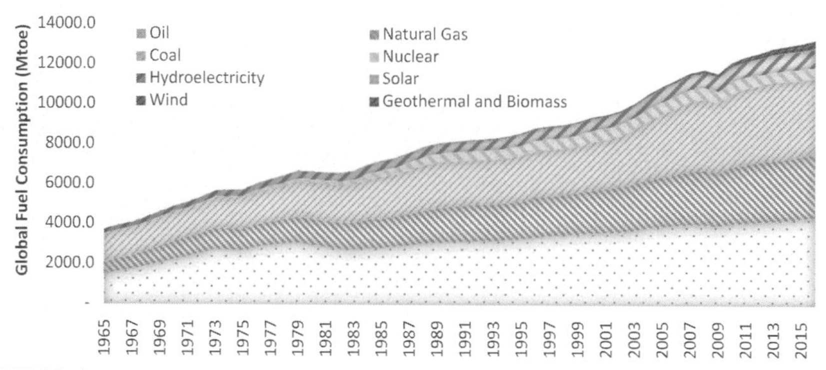

FIGURE 2.4

Historic fuel consumption in the world between 1965 and 2015.

Data from: BP Statistical Review of World Energy. 2017. BP, London.

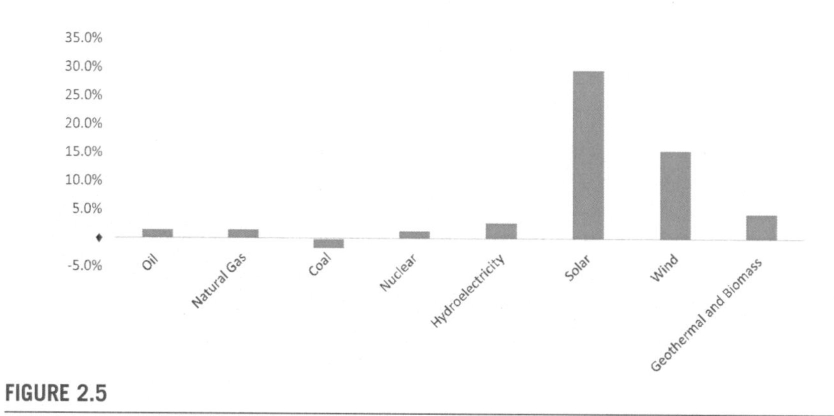

FIGURE 2.5

Growth rate per annum for the world's main energy sources for the year 2016.

Data from: BP Statistical Review of World Energy. 2017. BP, London.

concerning due to the environmental impact and the health concerns associated with these sources. In fact, the 3S+2S rule is very relevant in this case, as energy storage solutions are the primary key to success for the rising renewable energy sources such as solar and wind. Moreover, renewables such as solar and wind are growing rapidly despite their modest contribution to the world's energy mix. On the other hand, fossil fuels' growth is very limited while coal is actually going under depression as illustrated in Fig. 2.5. Therefore, solar is the fastest growing energy source in the world, followed by wind and geothermal.

2.2.1 Coal

This nonrenewable energy source has been the trigger of the industrial revolution and until now remains an important energy resource. In fact, it is the largest domestically produced source of energy in the United States and is used to produce significant amount of electricity. There have been some technological developments to make coal cleaner and more environmentally friendly. However, many of the coal plants have been shut down or converted into other energy sources such as natural gas because of their substantial environmental impact. Many chemicals can be formed when coal undergoes some processes such as gasification, refining, liquefaction, or coking. Furthermore, when using coal for electric generation, thermal efficiencies do not exceed 50%. In fact, further thermal efficiency improvements can be achieved by improved predrying and cooling technologies. An alternative to coal for electric generation is a derivative from coal cycle called, the integrated gasification combined cycle. Instead of burning coal upfront, the coal is gasified to create syngas, which is consequently burned in a gas turbine to produce electricity. Aside from environmental effects, coal imposes a health risk with numerous adverse health impacts and even deaths. Coal dust breathed by coal workers causes their lungs to turn black form its original pink color. Furthermore, the deadly London smog was created because of the excessive use of coal. There are different types of coal used for combustion including anthracite and bituminous, subbituminous, and lignite. Fig. 2.7 illustrates the recoverable coal reserves at the end of 2016. The figure clearly shows the abundance of coal resources still available. In fact, anthracite and bituminous coal composes approximately 45% of the total global coal, whereas subbituminous coal composes approximately 32% of the total available coal. Finally, lignite stands for only 23% of the total global coal.

Coal can be classified into two main categories as illustrated by Fig. 2.6. Hard coal such as anthracite can be used for domestic or industrial uses, whereas

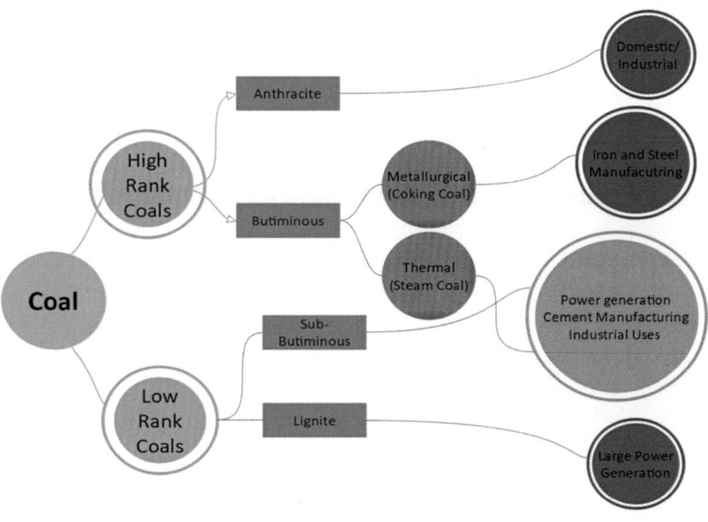

FIGURE 2.6

Classification of coal including various grades and corresponding applications.

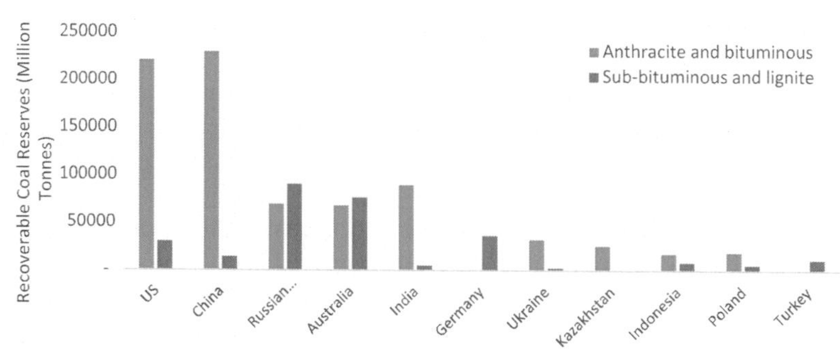

FIGURE 2.7

Recoverable coal reserves for the top reserves worldwide by the end of 2016 (mil tonnes).

Data from: BP Statistical Review of World Energy. 2017. BP, London.

bituminous is further processed through coking to be used in the iron and steel manufacturing processes. The differences in coal classification are also related to the cost, abundance, and environmental friendliness of each classification. Furthermore, the various processing techniques for coal also vary in their environmental impact and efficiency.

It can also undergo thermal steaming to generate power or be further used industrially. Moreover, low rank coals have high moisture content, yet low carbon content. Subbituminous is also used industrially and for power generation. Lignite is primarily used for large power generation, as it is the lowest coal grade. Despite the decrease of coal production and using petroleum or other fossil fuels, coal remains a popular source of energy especially for China and the United States. Fig. 2.4 shows 97% of the total coal production in 2011. It is evident that China is the biggest producer of coal in the world, where it accounts for approximately 50% of the total global coal production. In 2011, China alone produced 3520 million tonnes of coal. It is followed by the United States, India, and then the European Union consecutively. China's coal production started later than European countries. It took a steady and linear journey in the first half of the century by producing 179.7 million tonnes. However, in the second half of the century between 1950 and 2000, China's coal production skyrocketed and produced 3659.8 million tonnes. This was more than the production of the United Kingdom, Belgium, Germany, France, and Japan all together in these 50 years. The last two decades marked the peak production of coal in China. In fact, China scored the record in producing the highest amount of coal in one given year throughout the world. China's coal production took a big hit during the World War II (WWII) (1940s) and the Cultural Revolution (1960s). The American coal production has been undergoing a roller coaster effect throughout the years with a general direction toward increasing the production. Before 1890, the coal production doubled every decade, going from 6.9 in 1760 to 13 in 1860. After 1900, the production peaked in the year 1920 reaching 597.1 million tonnes of coal. After that, the increase in production followed a steep

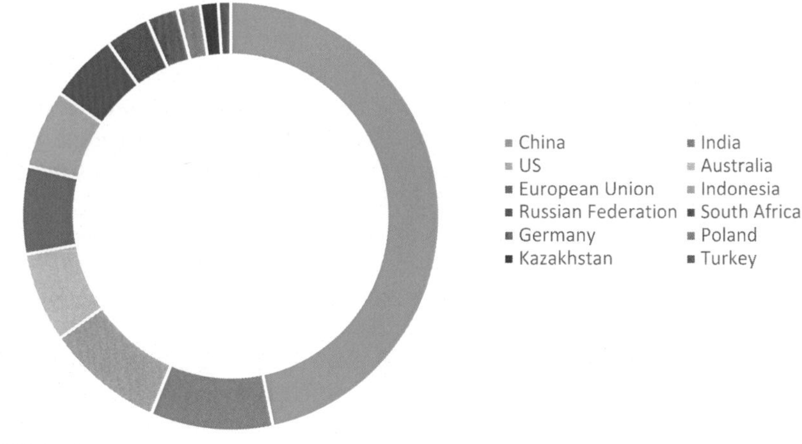

FIGURE 2.8

Coal production in 2016 and the top global producers.

Data from: BP Statistical Review of World Energy. 2017. BP, London.

linear trend from 1970 to 2000 reaching the highest annual production in American history in the year 2000 by producing 899.1 million tonnes. Fig. 2.8 shows the coal production in 2016 and the top global producers.

In 1800, the American produced almost 100,000 tonnes of soft coal, which was lower quality than anthracite. Anthracite sites were well known at the time close to Philadelphia, Boston, and New York; the problem was that there was not practical method to burn anthracite. The growth of the anthracite industry in 1820 in Pennsylvania led to the initiation of the American Industrial Revolution. Between 1700 and 1900, British coal production was almost equivalent to the total production of the rest of the world. Producing 881.6 million tonnes by 1900 was remarkable for the British Empire. As discussed before, this was a pivotal step in expanding the Empire and leading the industrial revolution. Until the mid-20th century, coal has maintained its utmost importance to the world by being the primary source of energy. Throughout the century, coal production increased, but its importance declined as the major production centers shifted. Furthermore, the world wars, relatively in close proximity of each other, had a huge impact on the world and on coal. World wars pushed the steel industry to peak their productions, thus peaking the coal production. After the wars ended, there were great shortages of coal that has led all affected national to maximize their coal productions.

Coal productions consequently met the demand quickly. In both wars, a large overcapacity developed causing the coal prices to drop significantly. After WWII, major events that contributed to ending the coal era were the switching of energy source for railroads and space heating. Railroads switched from steam to diesel, whereas space heating switched to natural gas and oil. In 1946, American railroads burnt 110 million tonnes of coal, whereas they only burnt 2 million tonnes in 1960.

Similarly, space heating of residential houses and commercial building dropped from 99 million tonnes in 1946 to 9 million tonnes in 1972. In this century, global coal trends varied. At first, Britain dropped from leading the world to becoming a minor coal producer. The United States took the British seat for a few years and was the biggest coal producer in the world. Shortly after, China became the world leader in coal production. Japan along with most west European countries reached their peak production around 1950s after which they declined tremendously. In East European countries, Russia, and South Korea, the peak in coal production was much later around 1980s.

Another substantial achievement in this century is the shifting from underground mining to surficial mining. At first, surficial mining only accounted for 10% of the American coal production in 1940. Within 60 years, surficial mining accounted for two-thirds of American coal. Over this century, different bodies of regulators have developed to control health and safety of coal mining. These organizations emerged in different names and independent of each other throughout the world.

Health and safety codes, acts, and regular inspections were developed to ensure the safety of the mine and the workers. Furthermore, the classification of coal in this century was more advanced as international trade of coal spread. Countries developed regional and national classification systems that they used. Coal gasification is also a process that was demonstrated by burning coal seam to produce gas that contains CO, CH_4, and H_2 and other high-order hydrocarbons. The complexities around this process aside from its economic infeasibility made it less attractive to investors.

Near the end of the century, coal producers consolidated, and by the year 2000, the largest 10 private companies controlled 23% of the world's production. Coal continued to be very competitive internationally and domestically. Fig. 2.9 shows

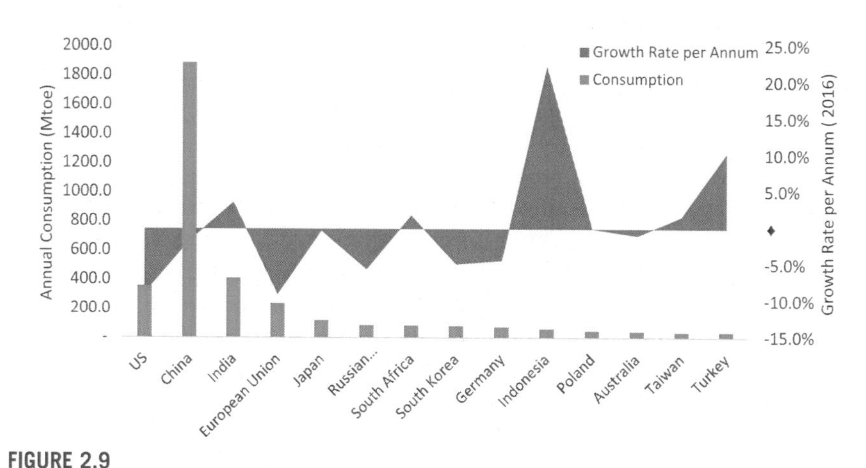

FIGURE 2.9

Coal consumption in 2016 for the top global consumers.

Data from: BP Statistical Review of World Energy. 2017. BP, London.

the annual consumption of coal for the top consuming countries worldwide in 2016 along with the growth rate per annum for each country. .

Coal is a unique mineral and has been the trigger for the industrial revolution. Coal resources are widely spread throughout the world and in massive amounts in table regions. The uses of coal vary from electricity generation to the production of gaseous and liquid fuels. These are all reasons that indicate that coal might still be valuable and a key player in the future. However, because of escalating regulations and restrictions on coal production, mining and use of coal will be restricted and thus more expensive. As nations respond to advocates of global warming and with the introduction of carbon tax, coal production will become expensive and might be less favorable than other current options such as natural gas. Natural gas is the main competitor for coal, as demand for electricity and energy is on the rise. On the other hand, a new player that also joined the competition is renewable energy.

Although natural gas is preferred over coal for being richer in hydrogen and poorer in carbon, renewable energy is the option that isolates carbon emission. The scientific community is heading in the direction of clean energy and providing viable energy solutions for societies with minimum environmental impact. These events indicate that coal might have little contribution to the world's energy demand. To have coal catch up, a life cycle assessment of coal production from mining to end use is ought to be investigated. Research is underway in minimizing methane (CH_4) release in coalmines. Methane, a GHG, is also the main cause of mine explosions. There are other environmental concerns associated with coal mining such as site restoration, management, and acid mine drainage. All in all, although coal has been criticized environmentally throughout its history, its growth was evident owing to its ludicrous characteristics such as low cost, ease of transportation, and its wide availability.

2.2.2 Oil, petroleum, and natural gas

Oil and natural gas have been dominating all aspects of our life, inheriting that from coal in the past century. It provides us with electricity, fuels our transportation systems, and is even used in the production of plastic goods and in fertilizers for our foods. These commodities have been extremely precious in our modern world and have been the key factors behind wars, world economics, and the current geopolitical map of the world as well as the transformation of social trends, lifestyle, and standards of living. In fact, crude oil is distilled and refined to obtain various types of fuels that are used for numerous uses as illustrated in Fig. 2.10.

Although we have capitalized on natural gas and oil in our modern world, we are not the first users. Earlier civilizations around 3000 BC such as the Egyptians and Babylonians relied heavily on oil. Crude oil that bubbled to the surface was utilized by these civilizations to waterproof their boats and as mortar in construction. Egyptians also used oil in preparing mummies to help preserve corpses. Herodotus, the ancient Greek historian, suggested that asphalt was used in the construction

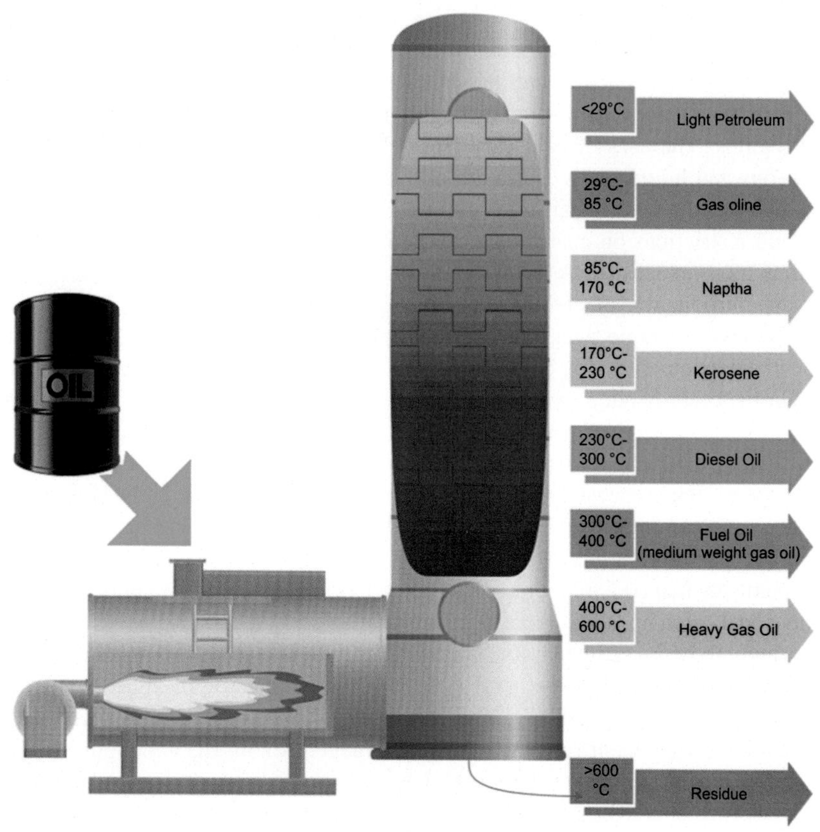

FIGURE 2.10

Oil distillation and the resulting products after refinery.

Babylonian towers around 2000 BC. Asphalt was recovered from neighboring riverbanks, as the oil described by Herodotus was natural leakage from the Jurassic Iraqi Petroleum. Ancient people who were mystified by the natural gas seepages sometimes built temples around these sites and worshipped the fire. Simply, these were natural seepages and gas ignitions from the ground. Around 625 BC in Babylon, asphalt was used for the first time as a road building material. Preserved Inscriptions archives the pavement of Procession Street, which extended from the palace to the northern wall of the city. The Chinese drilled the first oil wells in 347 AD that were 240 m deep and used bamboo poles to extract oil. It was considered a secondary commodity and used to evaporate brine to produce salt, which was considered the most important commodity in ancient times. Oil extract methods were tremendously primitive. Hand-dug wells at natural seep locations were pooled to a shallow depth. Oil was transported using camels, ship, and carts. Baku has gained the attention of numerous geographers, historians, and travelers from across the world.

Oil distillation took place until the 12th century when kerosene became available in Western Europe. Western exploitation expanded to North America where oil springs were found and documented by Joseph de la Roche D'Allion in 1632. Peter Kalm also published a map of Pennsylvanian oil seeps in 1753, whereas Moravian missionaries reported "oil skimming" and "oil seeps" from natural seeps. Moreover, the Seneca Indians also traded these products with the Niagara Peoples. In 1790, a man named Nathanial Carey from Titusville (the same town that hosted the first commercial petroleum well in the world in 1859) skimmed oil from natural seeps. He had collected it and delivered it to customers on horseback.

The Industrial Revolution as mentioned earlier was a paradigm shift in human history, as lifestyle and standard of living changed dramatically. The advancement in technologies in textiles, steam power, and metallurgy transformed the human sole reliance on physical labor, animals, and wind. With the ability of powering engines and manufactories, economic growth increased significantly compared with pre-Industrial Revolution when economic growth barely accommodated the population increase. Early in the Industrial Revolution, coal dominated the energy mix and was the fuel of choice, replacing wood and other preindustrial sources. Oil was used, however, in a limited capacity. Whale oil (rather than rock oil) was used for lubrication, heating, and lamp oil. Whale oil was favored as it produced a smokeless flame. Whale oil shaped the new light standard: the lumen. Consequently, whale oil became an important commodity as one single-sperm whale could hold almost three tonnes of high-quality sperm oil. In 1775, 45,000 barrels of sperm oil was produced annually by deploying 735 ships and hunting almost 2200 whales in New York, Connecticut, Rhode Island, and Massachusetts. In 1820, the United States was forced to import whale oils to meet the population demand of almost 5.3 million, of which approximately 1 million were slaves. It has been estimated that American whalers have killed more than 236,000 whales in the 19th century alone.

Natural gas was first used in America to fuel the street lamps of Baltimore, Maryland, in 1816. Shortly after, in 1821, William Hart dug the first successful natural gas well in Fredonia, New York, that was 27 feet deep, which is considered shallow, compared with today's standards. Following Rāzi's footsteps in the ninth century of distilling kerosene from petroleum, Abraham Gessner, a Canadian geologist, devised the technology to produce kerosene by refining bitumen, liquid fuel from coal, and oil shale in 1849. Kerosene replaced whale oil for lamp oil. In 1850, Gessner created his venture, called it Kerosene Gaslight Company, and installed lighting in the streets of Halifax. Four years later, he expanded his venture down south to the United States and created the North American Kerosene Gas Light Company at Long Island, New York. In 1848, F.N. Semyenov, a Russian engineer, drilled a well that was 21 m deep in Bibi-Heybat in Absheron. By 1861, this well accounted for 90% of the world's oil. Like Gessner, James Young, a Scottish chemist, also launched his venture in 1851 to produce paraffin by distilling coal and oil shale to produce kerosene. In 1853, Ignacy Lukasiewicz, a Polish pharmacist, improved Gessner refining method and produced clear kerosene from seep petroleum. A year later, Lukasiewicz launched the world's first modern "oil mine" in Bobrka with wells that were 30–50 m deep. Two years later, he launched the world's first industrial refinery at Ulaszowice. After Poland, Romania hosted the second modern commercial oil well in 1857, and it was the first country in the world to have its crude oil output recorded in international statistics at 275 tonnes. In the same year, Preston Barmore drilled two gas wells on Canadaway Creek in Fredonia, New York. He used 8-pound charge of gunpowder at a depth of 122 feet to "frack" the well. Barmore set the first record of artificial fracking. In 1855, Robert Bunsen invented the Bunsen burner, which mixed natural gas with air to provide heat for cooking and for heating space. James Miller Williams drilled the first oil well in North America at Oil Springs, Ontario, in 1858.

Colonel Edwin Drake triggered the modern oil industry by drilling the first American well using a steam engine, reaching the depth of 21 m at Oil Creek, Titusville, Pennsylvania in 1859. Initial production was 25 barrels a day, but by the end of the year, production dropped to 15 barrels a day. Although this event has evolved to be a milestone in the history of oil and natural gas, it had little novelty. Perhaps Drake's well featured the use of the steam engine for drilling compared to digging of previous wells along with organized commercial effort that supported the project and the fact that the site was a major site. On the other hand, Lukasiewicz's wells in Poland were deeper than Drake's by more than 10 m. There was also considerable activity since the mid-19th century worldwide. Engine-drilled wills in West Virginia were established in the same year as Drake drilled his well. The world first modern wells and refineries were already established in Poland and Romania as mentioned earlier. Fig. 2.11 shows the world proved oil reserves as of 2016, including the countries with higher than 30 thousand million barrels. As observed, Venezuela in addition to Middle Eastern countries has the highest oil reserves in the world. In fact, the Middle East houses more than 48% of the world's oil reserves.

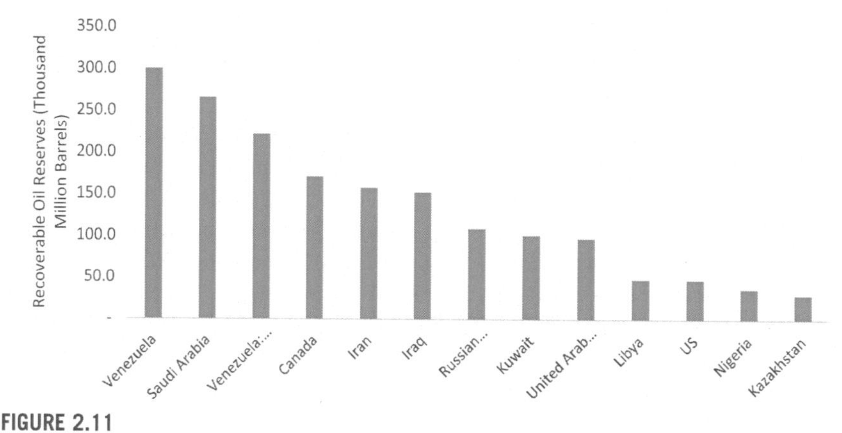

FIGURE 2.11

Recoverable oil reserves for the top reserves worldwide by the end of 2016.

Data from: BP Statistical Review of World Energy. 2017. BP, London.

By 1860, Gessner and competitors established 40 kerosene plants across North America. John D Rockefeller & Samuel Andrews competed against Gessner and created Standard Oil. America's oil production that year was 0.5 million barrels, which quadrupled in the following year reaching 2.1 million barrels in 1861. By 1870, the average American or British citizen consumed 2.2–2.5 tonnes of oil annually. This is relatively equivalent to the average person's consumption in modern Senegal or Haiti. The first oil drilling in Venezuela was in 1878 at Lake Maracaibo.

Kerosene cost in 1870 dropped to 26 cents per gallon after its peak cost of 59 cents per gallon in 1865, after the Civil War. In this year, Standard Oil led by John D Rockefeller was incorporated in Ohio and had destroyed all its competitors in Cleveland. In the next few years, Standard Oil spreads throughout the United States and abolishes all competitors by dominating the northeastern United States. By 1877, Standard Oil controlled 90% of American refineries. Standard Oil remained a monopoly as the world's largest oil refiner until its breakup in 1911 by Supreme Court. This made John D Rockefeller the richest man in the world as he was worth the equivalent of $318 Billion (2007 dollars). Today, Bill Gates is worth in the order of $85.7 Billion. Even when Standard Oil broke up, Rockefeller profited a lot from the settlements. Rockefeller retired in 1895.

The 19th century featured a community of inventors, engineers, scientists, and revivers who led to the lifestyle transformation of human civilizations. The constant invention and improvement of new technologies that reduced the reliance on manual labor or animal power were impactful. Regardless of how much emphasis there is on this transformation, the radical nature of this paradigm shift cannot be overlooked. In 1876, Nikolaus Otto, working with Gottlieb Daimler and Wilhelm Maybach, started building the four-cycle engine. The year 1879 was a remarkable milestone in human history as Thomas Edison invented the electric light bulb. In the same year, Karl Benz patented his internal two-stroke combustion engine. The world's first car was developed by Benz in 1885, which featured a four-stroke engine and ran using gasoline, which is a cheap by-product of kerosene. Natural gas can also be found in abundance in Asia as illustrated in Fig. 2.12.

In 1893, Rudolf Diesel patented his first compression ignition (diesel) engine. Two years later, the invention of combustion engine takes place. Henry Ford's first motorcar came on the streets in 1896. Henry Ford fulfilled his promise to create a car that anyone could afford with the model T in 1908. This increased the demand for gasoline dramatically. By 1900, there were 8000 registered automobiles in the United States. In only 20 years, there were 8.5 million vehicles registered. January 10, 1901, marked the birth of modern oil industry. The eruption of crude oil also known as black gold at the Spindletop oil field in Pennsylvania changed the world forever. Drilling continued reaching the depth of 1020 feet after overcoming initial drilling obstacles due to the sandy ground. Upon eruption, oil reached a height of 150 feet, producing close to 100,000 barrels a day. This event caused the oil prices to go down from $2 to $0.3 per barrel. Many major oil companies in America such as Exxon, Gulf Oil, and Texaco trace their origins to this site. Within months of this event, tens of thousands of Americans assembled in Spindletop and transformed

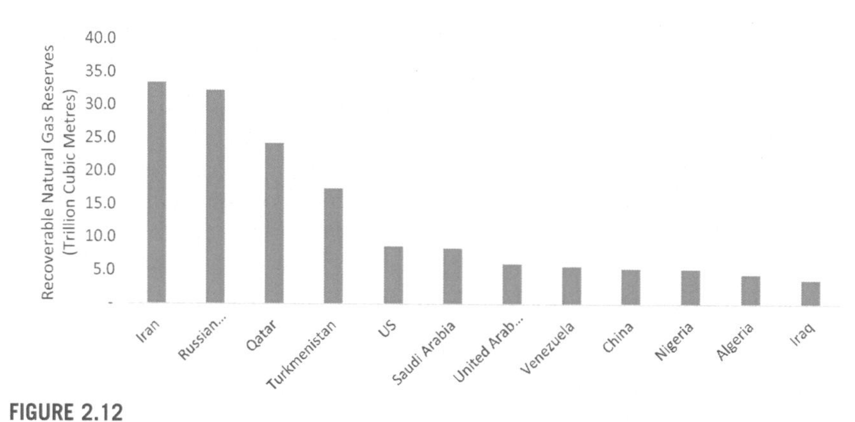

FIGURE 2.12

Recoverable natural gas reserves for the top reserves worldwide by the end of 2016.

Data from: BP Statistical Review of World Energy. 2017. BP, London.

the southeastern Texas from a farmland to a boomtown within months. In the first year, this site produced more than 3.5 million barrels. In the second year, it produced more than 17.4 million barrels. This site destroyed the monopoly established by Rockefeller and Standard Oil and urged oil prices to go down. The abundance of oil around Texas had major social impacts. Major oil companies and refineries along with marketing companies all trace back to this event. Thousands of new jobs were created and the economy flourished. The availability of oil inspired further technological advancements to take place in the United States. On December 17, 1903, the Wright brothers piloted the first powered airplane 20 feet above a North Carolina beach. They flew for 12 s, covering 120 feet. Three other flights took place that day with a record of 59 s and covering 852 feet. Although the idea of aviation started with Abbas Ibn Firnas in the seventh century, this key landmark was a significant highlight in the history of oil and natural gas as it introduced a new method of transportation to society that will later grow to be a revolutionary method that is widely used worldwide. In the same year, Ford Motor Company was founded. The exponential growth of oil and natural gas resulted in a standard of living that is extremely dependent on hydrocarbon energy. Top global oil producers in 2016 include the United States, Saudi Arabia, and Russia as indicated in Fig. 2.13. Furthermore, many Middle Eastern countries along with some South American countries are among the world's top oil producers.

In 1901, Englishman William D'Arcy had received a license to explore for oil in Persia. He deployed George Reynolds who searched for 7 year with no results. On May 26, 1908, the last attempt to drill has been changing the globe in all aspects. At 4 in the morning, the drill had reached 1180 feet below the desert sands as a huge gusher shot 75 feet into the air. The first Middle Eastern oil well has been successfully established. Although Arabs did not have the knowledge or technology to extract oil, Western companies rushed to Masjid Suleiman in order to secure oil exploration and extraction. The Middle East provides 60% of the world's oil supply.

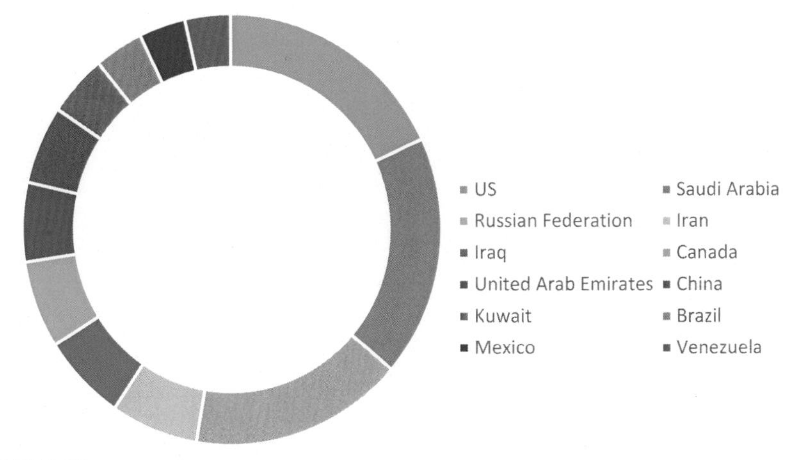

FIGURE 2.13

Top global oil producers in 2016 (1000 barrels/day).

Data from: BP Statistical Review of World Energy. 2017. BP, London.

Ironically, at the time the geologic consensus was of the opinion that there was no oil in the Arabian Peninsula. A few years before this event, Abd-al'-Aziz ibn 'Abd al-Rahman Al Sa'ud stormed Riyadh, taking the city from the Rashid tribe and later establishing the Kingdom of Saudi Arabia. In 1907, the Knickerbocker Crisis marked the fall of the New York Stock Exchange by 50%, but the financial panic quickly recovered as significant oil fields have been discovered in Canada, Sumatra, Iran, Venezuela, and Mexico, all of which were being developed at an industrial level. After the discovery of oil in the Middle East, Anglo Persian Oil Company was formed and later changed to British Petroleum. The year 1910 was eventful as the US Congress authorized a legislation that allows setting aside land under the category of Naval Petroleum Reserves. In 1912, Elk Hills and Buena Vista Hills were the first two Naval Petroleum Reserves. Back to 1910, Lakeview gusher blew out near Los Angeles, California. The Lakeview gusher is considered America's greatest gusher. Because there are other gushers that are thought to be the biggest, the following are the specifics of the top three:

- The Spindletop gusher: spilled 900,000 barrels for 9 days at a rate of 100,000 barrel/day.
- The Gulf oil spill: 4.9 million barrels leaked throughout a period of 65 days. The peak rate reached 62,000 barrels/day. 185 million gallons of oil was spilled.
- The Lakeview gusher: spilled 9.4 million barrels that leaked for 18 months with a peak rate of 125,000 barrels/day/. 395 million gallons were spilled.

Along the same lines, in 1919, Hay No. 7 in Elk Hills catches in fire and becomes the greatest gas gusher in America. At the US Naval Petroleum Reserve, Nay No. 7 blew out and caught on fire on July 26, 1919. It flew 50 million cubic feet (MMCF)

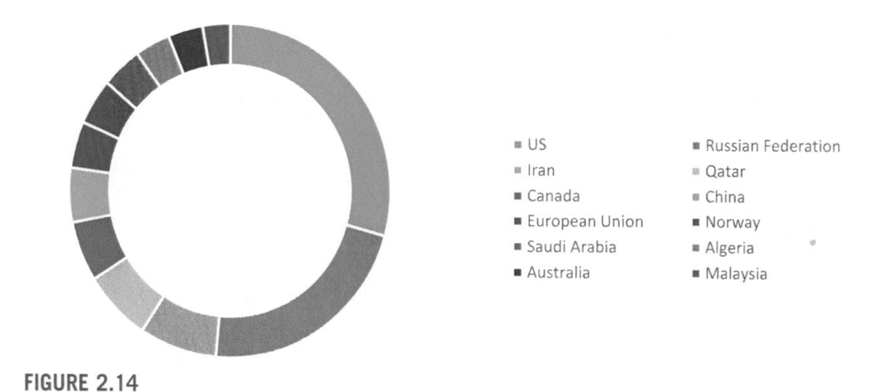

FIGURE 2.14

Top global natural gas producers in 2016 (1000 barrels/day).

Data from: BP Statistical Review of World Energy. 2017. BP, London.

of gas daily. Burning continued for 26 days until extinguished by dynamite. This site produced 43 billion cubic feet of gas in the following 7 years. Natural gas production in 2016 was dominated by the United States and Russia with over 500 Mtoe each as illustrated in Fig. 2.14.

With the escalation of the World War I, oil and natural gas proved to be a critically important asset for the protection of nations. Naval warships, trucks, and tanks all depended on oil. Since the British Empire had few oil resources, it became strategically more ludicrous to control the Middle Eastern oil. The use of seismic waves as a tool to explore the Earth's interior was introduced shortly after the war between 1921 and 24. Experiments were conducted in southern United States. This key development made exploration much easier and more readily available.

Although the great depression was very harsh for many societies across the world, the advancement of energy production and innovation did not decline. In fact, it was only 2 years later that oil was discovered in Bahrain in 1932. Frank Holmes leased a concession from the Sheikh of Bahrain and with the aid of Standard Oil of California; they drilled and extracted massive amounts of oil. Furthermore, the following year in 1933, Standard Oil of California (later became Chevron) created a subsidiary company and named it California Arabian Standard Oil Corporation, which was also granted drilling rights by Saudi Arabia. This later became the famous Saudi Aramco. Today, Aramco is the world's largest oil company, producing more than 260 bnbbls of reserves, which equals approximately 300 trillion cubic feet of gas reserves. Moreover, it produces more than 8 billion barrels annually and operates the world's largest hydrocarbon network. Shortly after these significant events, the world's second largest oil field is discovered in Burgan, Kuwait, in 1938. In the same year, Mexico nationalizes foreign oil companies, where all assets were put under the management of Pemex (Mexican Petroleum). Fig. 2.15 shows the major oil consumers worldwide along with the growth rate per annum in their annual consumption. The United States, European Union, and China are considered the

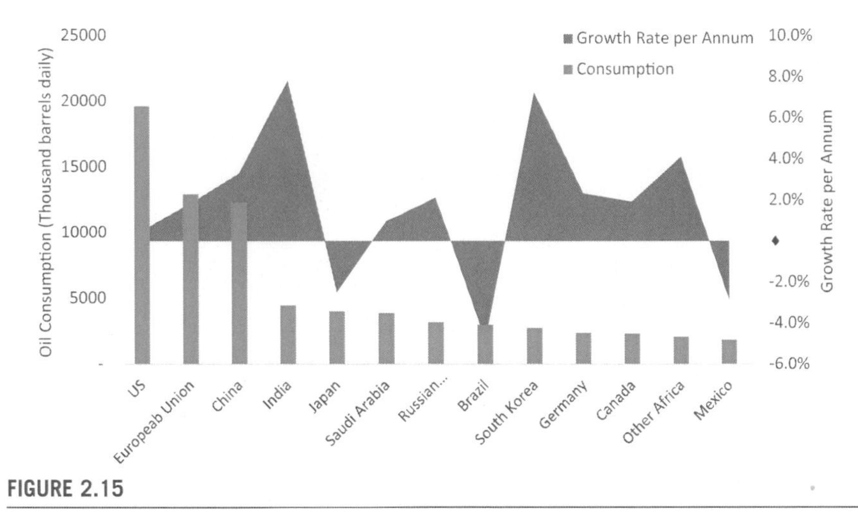

FIGURE 2.15

Oil consumption in 2016 for the top global consumers.

Data from: BP Statistical Review of World Energy. 2017. BP, London.

heaviest consumers of oil in the world. However, India and South Korea are considered the countries with the highest growth rate in oil consumption per annum.

The WWII struck the globe in 1939 as the most industrial, mechanized, and fierce war in human warfare history. The industrial importance in this war gave more value and priority for oil. In fact, oil was a strategic and critical asset during the war. The Allies strategically targeted oil refineries, storage depots, and oil fields, as they considered them the lifeblood for Germany. The Middle Eastern oil along with the famous Baku oil played a crucial part in this World War, as they supplied the oil needed to run the events and affairs of this war. The isolation of Japan from its oil supplies caused it to be considerably weaker near the end of the war. Furthermore, the Axis powers considered Baku and the Azeri oil fields as a strategic target.

In 1955, Egypt nationalized the Suez Canal, which caused the Suez Canal Crisis a year after. The Canal was the primary route of British oil coming from the Middle East to Britain. After the Egyptians nationalized the canal, they quickly renegotiated the terms of use with the British to get their fair share of the oil profits. This decade witnessed a number of events such as the discovery of natural gas in Netherlands, oil discovery in Nigeria and Algeria. Furthermore, OPEC (Organization of Petroleum Exporting Countries) was founded in 1960 in Baghdad, which included major oil exporters such as Saudi Arabia, Kuwait, Iran, Iraq, and Venezuela with the intent of coordinating and influencing the oil market in the world.

The largest single oil spill took place in Bay of Campeche in the Gulf of Mexico in June 1979 and was brought under control in March 1980. Between 1979 and 1981, oil prices rise from $13.00 to $34.00. In 1984, Chevron acquires Gulf Oil after a bidding combat with Arco. In 1997, while Qatar celebrated its inauguration of establishing the world's first significant liquid natural gas exporting facility, the Koyoto

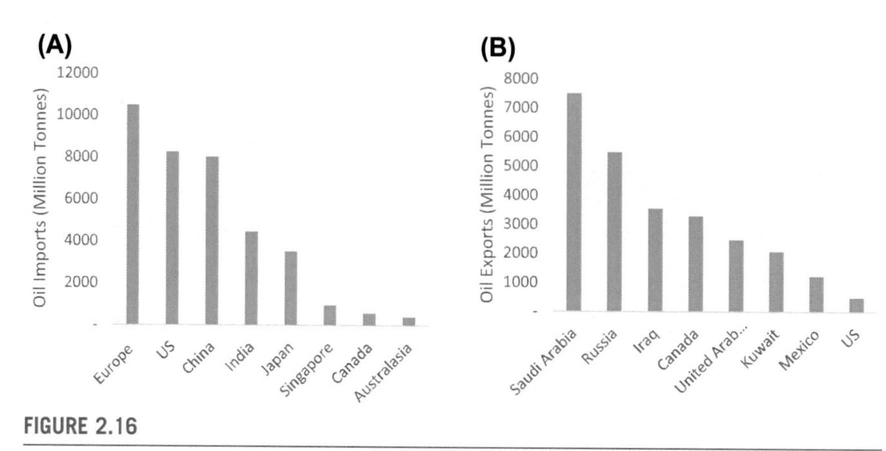

FIGURE 2.16

Crude oil (A) imports and (B) exports in 2016 for the top global players.

Data from: BP Statistical Review of World Energy. 2017. BP, London.

Agreement passed to limit the production of GHGs. A year following the Koyoto Agreement, however, the United States started its fracking revolution. Fracking is a novel method that was used to extract oil by pumping high-pressure fluids into ground fractures, which allows the oil to be extracted easily. This method brought down US oil imports significantly, while the British were still exploring the fracking option. Fracking definitely influenced the oil economy positively in the United States and Europe. The demand on energy and oil kept increasing while modern civilization went on to shape its identity. Crude oil imports and exports are very dynamic because of various associated factors. Fig. 2.16 shows the top global importers and exporters of crude oil in 2016. The major importers of oil include Europe, the United States, China, and India, whereas the major exporters of oil are Saudi Arabia, Russia, Iraq, and Canada.

In summary, oil and natural gas have evolved to be an essential energy resource of the modern world. In other words, humans on the earth in 2017 have developed a lifestyle over the past few centuries where oil is indispensable. Oil is embedded in every aspect of our modern lifestyle as it comes in many different shapes or forms. Let us analyze an average day of a modern person on the earth in 2017. One would wake up in the morning and turn off the plastic alarm clock, which was made from oil by-products to taking a shower using a shower gel and a showerhead. An average person in this modern world then uses a bath mat to dry and brushes their teeth using a toothpaste and a toothbrush. The very food we eat from the fridge in the morning such as cereal would has artificial fertilizers to the very morning routine one might have from watching TV, browsing a magazine, or pouring milk from a milk carton. All these luxuries are accessible thanks to oil. Then one goes to driving their car, which burns fuel and uses car tires to transport the vehicle on asphalt. Oil appears vividly in every aspect of modern life and comes in many shapes and forms. Furthermore, the detergent used in mobbing, the water bottle carried around, the takeaway cup are all forms of oil's influence over our society's lifestyle. Oil have revolutionized human history once again after the first industrial revolution caused by coal. The recent economic dynamics, political agendas, wars, geography of nations, social life, and cultures have all been directly impacted by oil. Therefore, it is crucial that humans understand the external impacts and factors that shape their life.

2.3 Nuclear energy

Nuclear energy is simply the use of nuclear reactions, which release nuclear energy that is further processed to generate heat to power steam turbines for the purpose of electricity generation. Nuclear power can be obtained through three different methods: nuclear fission, nuclear fusion, and nuclear decay. Nuclear fission of specific elements is currently the most popular method. Some consider nuclear energy as clean energy because of its low carbon power generation. In fact, nuclear energy's GHG emissions as comparable with renewable energy in terms of total life cycle emission per unit of generated energy. Although this method of power generation is ludicrous, there are only 449 operable fission reactors in the world as of 2018. This is because of its hazardous nature and possible deployment in dangerous activities. However, nuclear energy's electrical capacity exceeds 394 GW as it accounts for more than 10% of the global electricity generation in 2012 (IEA, 2018). Furthermore, although nuclear energy is more environmentally friendly than fossil fuels, social debate about their safety, sustainability, and waste disposal remains heated. Such incidents indeed occurred in Generation I and II reactor designs, including the Chernobyl disaster in 1986, the Fukushima Daiichi disaster in 2011, and the Three Mile Island accident in 1979.

The term "atom" was first introduced by the Greek philosopher Democritus of Abdera, and he defined it as smallest constitutes of matter. Atom is therefore a Greek word in origin, which means "not divisible." However, after the introduction of nuclear fission, the splitting of the atom has become possible for energy harvesting. Uranium was discovered in 1789 by a German chemist, Martin Klaproth. It was originally named after the planet Uranus. A decade later in 1803, John Dalton, a British chemist stated that elements form because of certain atom combinations. He also stated that atoms of the same element were all identical. This foundation kept scientists busy as they concentrated on the classification and identification of elements as well as arranging them in an organized and unique fashion. Various scientists attempted to rearrange the elements until we obtained the current Periodic Table. Wilhelm Rontgen discovered the X-rays in 1895 by passing electric current through a glass tube. Rontgen had sucked the air out of the glass tube and was experimenting with the cathode rays. In his experiment, he had the device covered, yet noticed that there was light coming from the side photographic plates when the device was energized. Rontgen quickly realized that this is a new kind of ray and called it X-ray. Two weeks later and after continuous study of these rays, he took the first X-ray, which was a photo of his wife's hand. Rontgen therefore became the father of modern medical diagnostics. Soon after this discovery of the X-ray, Becquerel discovered that spontaneous rays emitted by uranium salts. In 1896, Becquerel left uranium salts sitting on photographic plates and noticed that they would expose although the tube containing the cathode ray was not energized. He hypothesized that the energy must have come from within the salts. After studying this phenomenon in detail in 1898, Marie Curie and her husband Pierre discovered and added two new elements to the Periodic Table, polonium and radium. They branded this phenomenon as radioactivity. Rutherford and Soddy investigated this phenomenon further and came to understand that elements emit three different types of radiations: alpha, beta, and gamma. Ernest Rutherford dominates the next two decades and advances the understanding of radiation. Max Planck, a German scientist in 1900, introduced Planck law that states that energy of each quantum is equivalent to the electromagnetic radiation frequency multiplied by his constant, the Planck's constant (h). Planck discovered this universal constant as he stated that energy is emitted in small individual units, he called quantum. This discovery was the birth of a new scientific field called quantum mechanics and opened the door for further research in the field of nuclear energy. Einstein suggested that the energy and mass were interrelated by the speed of light. Rutherford discovered half-life and that the majority of the element's mass is focused in a tiny nucleus. Fig. 2.17 shows the different generations of nuclear reactors with the various technologies used under each generation.

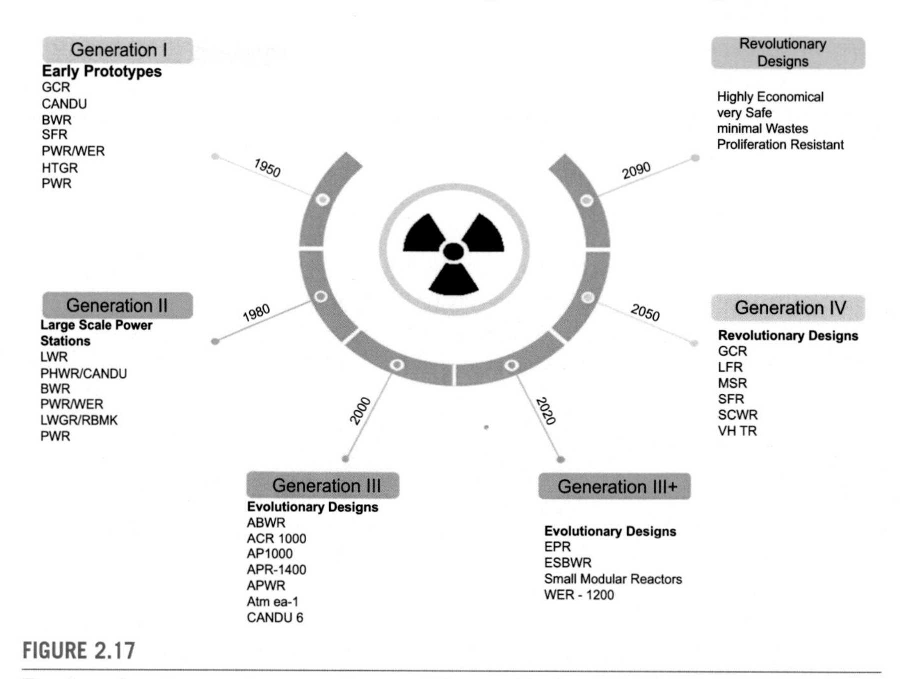

FIGURE 2.17

Timeline of nuclear reactor generations and the used technologies for each generation.
Adapted from Ricotti, M. E., 2013. Nuclear energy: basics, present, future. EPJ Web of Conferences 54, 1005. https://doi.org/10.1051/epjconf/20135401005.

Later on in 1920, he theorized that neutrons exist as neutral particles in the nucleus. At that time, there was no evidence that neutrons existed yet. It was not until a decade later when Chadwick identified the neurons in 1932.

Rutherford's atomic model as he anticipated was composed of a positive central region where the mass was concentrated and electrons revolved around this center in orbits, similar to our solar system. Enrico Fermi performed experimentations in 1934 that demonstrated how neutrons were able to split many kinds of atoms. Fermi experimented with uranium and was surprised to find that the resulting elements had a much lighter atomic mass than uranium. Near the end of 1938, Fritz Strassmann and Otto Hahn demonstrated that barium was the new lighter element, which was the leftover after using uranium. Their experiment proved that atomic fission was possible and had occurred. This was later explained by Lise Meitner and Otto Frisch when they suggested that the neutron was captured by the nucleus, which led to intense vibrations causing the nucleus to split into two unequal parts.

Bohr visited the United States in 1939 and shared the findings of Strassmann and Hahn with Einstein. They discussed the possibility of a self-sustaining chain reaction. The scientific community started to accept the idea of a self-sustaining chain reaction with the right amount and configuration of uranium. In that year, Fermi and Szilard measured neutron multiplication and concluded that a nuclear chain reaction is indeed possible. Whereas Einstein, Wigner, Teller, and Szilard sign a letter warning President Roosevelt of the possibility of nuclear weapons/warfare, Roosevelt authorizes the establishment of the Advisory Committee on uranium; thus starting the American nuclear bomb effort. However, these efforts were not progressive or vigorous.

Fermi led a group of scientists in 1942 at the University of Chicago to develop their theories around the first nuclear reactor. Later that year, they commenced the construction of Chicago Pile-1 (CP-1), the world's first nuclear reactor. The world entered the nuclear age on exactly December 2, 1942, when Fermi and his group initiated the first demonstration of CP-1, where the reaction became self-sustaining. This scientific group were the first to transfer these theoretical scientific ideas into a real technological success. After this success, the Manhattan project launched with the objective of assembling the atomic bomb. Filled with ultimate secrecy, secret cities were established in Oak Ridge to enrich uranium, Hanford to produce plutonium and in Los Alamos to design and assemble the bomb.

The first atomic bomb hit the ground for testing on July 16, 1945, in the desert of Alamogordo. It was an evident success, which made it serviceable in the WWII. Two atomic bombs were dropped on Japanese cities Nagasaki and Hiroshima on August 6 and 9, 1945. The bombs killed more than 240,000 people, and their effects are evident until today. This atrocious incident ended the WWII by the unconditional surrender of the Japanese. The top nuclear consumers in the world in 2016 include the United States, European Union, France, China, and Russia as illustrated in Fig. 2.18. The strategic success of nuclear power in the WWII encouraged the American Congress to support the development of nuclear energy for peaceful and civilian uses in 1946. They created the Atomic Energy Commission, which authorized the

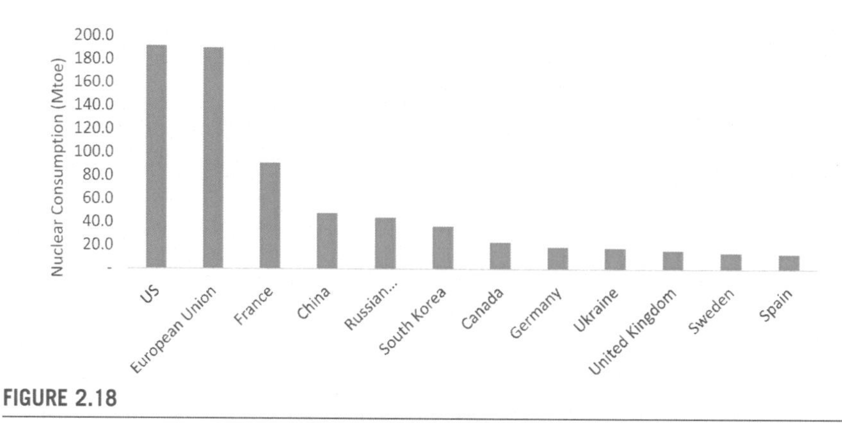

FIGURE 2.18

Top nuclear consumers in the world in 2016.

Data from: BP Statistical Review of World Energy. 2017. BP, London.

construction of Experimental Breeder Reactor I in Idaho. Electricity was generated for the first time from this reactor on December 20, 1951. The 1950s was generally known for the marketing of nuclear energy as a plausible and profitable source to produce electricity. The first commercial electricity-generating plant using nuclear energy reached its optimal design power in 1957 in Shippingport. They used ordinary water to cool the core of the reactor during the chain reaction. Light-water reactors were the best nuclear reactor designs to date.

The first soviet atomic bomb was constructed in 1952 with the design of a cylindrical container containing the atomic bomb on one end and the hydrogen fuel on the other. Hydrogen pressure is ignited by the amount of radiation upon the outbreak of the atomic bomb. The atomic bomb is found to be 700 times higher than that of Hiroshima's atomic bomb.

In 1953, Eisenhower launched a civilian program through his Atoms for Peace speech. Furthermore, the United States turned to the United Nations to denounce terror in the world's population. They also delivered a warning that had the United States been attacked by nuclear weapons, the attacker would be destroyed. The Soviets continued to expand their nuclear program and the first nuclear-powered submarine was launched, as the Obninsk reactor became the first commercial nuclear power plant in 1954. A number of international organizations were created amid this cold war to contain the atomic energy and to encourage safe and civilian uses instead of military ones. The International Atomic Energy Agency (IAEA) was founded in 1957 in Vienna as well as the Nuclear Energy Agency (NEA) based in Paris. In 1956, the British launched their nuclear program by establishing the first nuclear power station. General Electric was the pioneer in receiving commissions to build a strictly commercial power plant in 1963.

After this event, requests to build nuclear power plants for industrial and commercial purposes hailed significantly. Fig. 2.19 displays the growth rate per annum and the consumption of the top nuclear countries in the world.

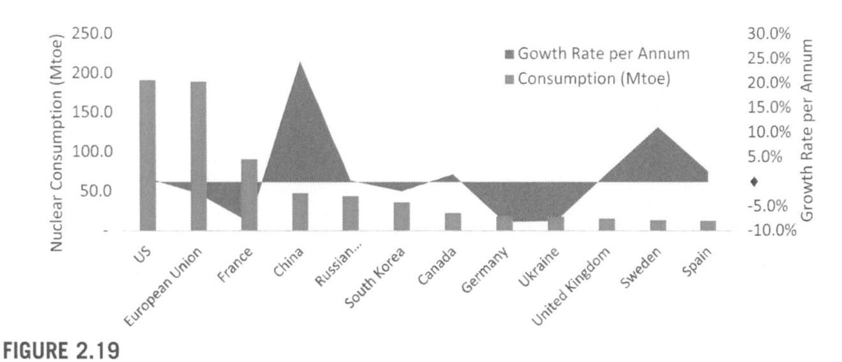

FIGURE 2.19

Nuclear consumption in 2016 for the top global consumers along with the growth rate per annum.

Data from: BP Statistical Review of World Energy. 2017. BP, London.

In 1967, a decade after the establishment of IAEA, the Nuclear Nonproliferation Treaty is recognized and signed by many countries across the world. The participants of this treaty agreed not to transfer nuclear weapons or work for manufacturing and pledged to figure out the necessary safeguards to comply with this treaty. The following were the main treaties signed:

- **Antarctic Treaty**: 37 countries pledged to avoid using this territory for any nuclear explosions, disposal, or radioactive waste.
- **Ban Treaty Nuclear Weapons Tests in the atmosphere and in outer space and underwater**: the United States, the United Kingdom, and the former USSR act as repositories.
- **Treaty "Principles Governing the Activities of States in the Exploration of Outer Space"** includes the United States, the United Kingdom, and the former USSR act as repositories and pledge to not expose the moon and other celestial bodies to any nuclear weapons.
- **Ban Treaty Nuclear Weapons in Latin America**: Mexico, signed in 1967.
- **Nuclear Nonproliferation Treaty**: In force since 1972 and extended in 1995 with the United Kingdom, the United States, and the former USSR as depositories.

The French Prime Minister Messmer launched a massive nuclear program to counteract the oil crisis. This had a long-term on the French energy composition. In 2004, nuclear energy accounted for 75% of the French electricity. In 1979, Three Mile Island reactor suffered a partial meltdown; however, radiation was largely contained. In 1986, EBR-II showed that advanced, sodium-cooled reactors are able to shut down passively without a backup system. In that same year, a major disaster in the history of nuclear energy takes place. The Chernobyl reactor suffered a large power excursion that caused the release of huge amounts of radiation. This incident led to the death of more than 50 firefighters, and it was estimated that at least 4000 civilians would die of early cancer. All in all, nuclear energy remains a popular

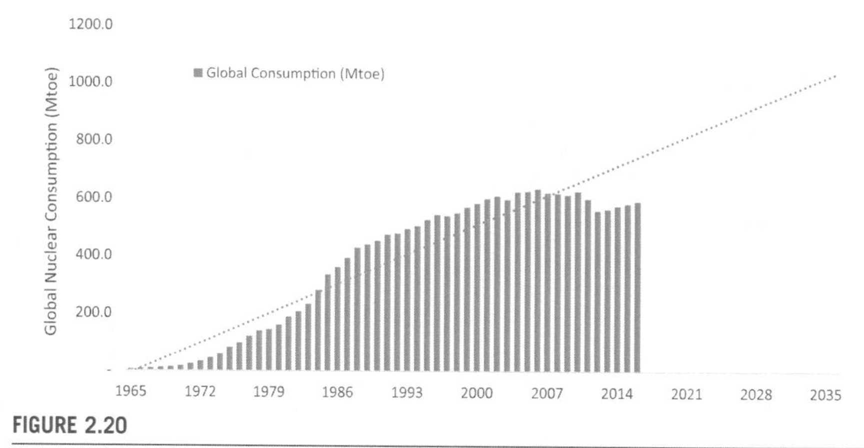

FIGURE 2.20

Global nuclear consumption between 1965 and 2016.

Data from: BP Statistical Review of World Energy. 2017. BP, London.

source of energy with steady growth worldwide. Fig. 2.20 shows the global nuclear consumption from 1965 until 2015 along with a forecasted growth in consumption based on a linear trend.

2.4 Renewable energy

Alternative and renewable energy emerged long time ago for many reasons such as availability and simplicity. For example, wind turbines were as a primary energy source and then later transformed into mechanical energy in European farms. However, these sources of energy were not considered commercially viable and were not deployed at large scale until recently due to rising global environmental conscious. This section will dwell more on various renewable energy sources such as wind, solar, geothermal, and hydrogen.

2.4.1 Wind

Wind power is the utilization of airflow to provide mechanical power used for electric generation. It is considered a clean and a renewable energy source. In fact, wind power is a form of solar energy because winds are created by the uneven heating of the atmosphere by the sun, the earth's rotation, and the irregularities of the earth's surface. Furthermore, wind power is considered environmentally friendly as its operation emit zero GHGs, does not consume any water, and uses little land size. Wind farms are a collection of individual wind turbines, which are interconnected to an electric power transmission network. Wind farms are established both onshore and offshore. These wind farms could feed the grid and support the local demand or support small off-grid communities. Furthermore, wind power is an intermittent source of energy, meaning that it is not available all the time. The energy derived from the wind is in the form of kinetic energy. The total wind energy flowing through an imaginary surface with area A during time t is:

$$E = \frac{1}{2}mv^2 = \frac{1}{2}(Avt\rho)v^2 = \frac{1}{2}At\rho v^3 \tag{2.1}$$

where Avt is the volume of air passing through A, ρ is the air density, and v is the wind speed. All wind turbines are designed to accommodate various wind speeds.

Fig. 2.21 shows various types of wind turbines along with their respective prototypes and applications. Most turbines' cut-in speeds are between 3 and 4 m/s, whereas the cut-out speeds are around 25 m/s. However, survival rates are designed much higher. Survival speed rates are the maximum wind speeds that a turbine can handle, above which the turbine may malfunction. The survival speeds range from 40 to 72 m/s. Furthermore, Table 2.1 shows the specifications of the main components of a wind turbine.

Wind power is capital intensive but has no fuel costs. Furthermore, this energy source has reached grid parity, meaning that the wind power costs match traditional energy sources for some parts of Europe and the United States. However, renewable energy sources usually grow after a period of government incentives such as feed-in-tariff. The advantages of wind power are that it is a free, clean, nonpolluting, and renewable resource. Fig. 2.22 shows the historic consumption of energy from wind between 1978 and 2015 with projected consumption for the next decade. Although wind energy has a modest contribution to the global energy consumption mix, its use is increasing exponentially. This is due to the rising maturity of the technology as well as the economic viability and the technical and feasibility success coupled with this energy source.

On the other hand, it is costly at first, as it requires significant initial investment than conventional energy generation. In addition, environmental concerns include

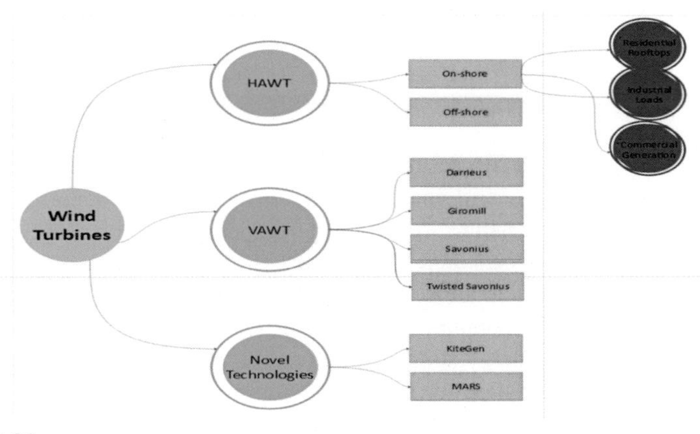

FIGURE 2.21

Classification of wind turbines and corresponding applications.

Table 2.1 Main components of a wind turbine along with description for each component.

Component	Description
Rotor blades	Three blades; blade tips speed/wind ratio of 6–7
Gear box	Converts 15–20 rotations per minute to 1800 revolutions per minute for electric generation
Nacelle	Houses the gearbox and generator connecting tower and rotor
Generator	Electric generator to convert mechanical energy into electricity
Power cables	Cables connecting the generator to the transformer
Tower	Higher altitudes provide higher velocities due to aerodynamic drag and air viscosity
Transformer	Converts the incoming generated electricity to the right voltage for before feeding into the electricity grid
Grid	Existing interconnected network for delivering electricity

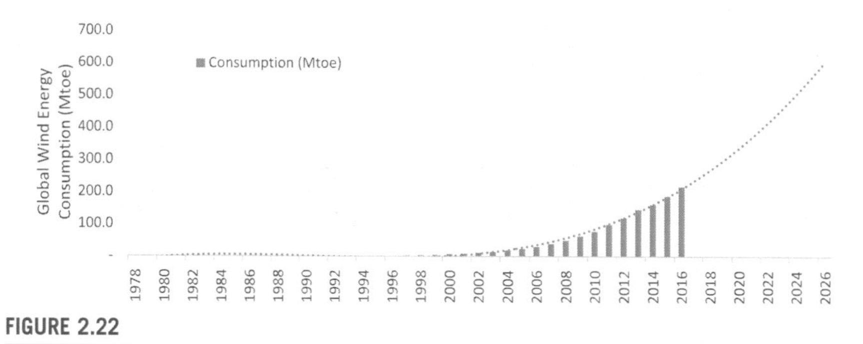

FIGURE 2.22

Global wind energy consumption between 1978 and 2016 with projected consumption for the next decade.

Data from: BP Statistical Review of World Energy. 2017. BP, London.

noise pollution from the blades, aesthetic visual impact as well as interference with local wildlife. For example, bird and bats killed by driving into the rotor. Finally, because of its intermittent nature, supply and transport issues are also a concern. In fact, wind energy cannot be stored unless batteries or a secondary energy resource such as water are used, which are seldom. In conclusion, for this energy source to prosper and become more popular, social acceptance and public awareness must be integrated along with energy storage solutions. Fig. 2.23 shows the growth rate per annum and the consumption for the world's top wind energy consumers.

It is evident that Brazil, India, and China express the highest rates of growth in 2016, whereas the European Union, China, and the United States dominate the wind energy consumption in the world. Furthermore, Fig. 2.24 illustrates the rate at which

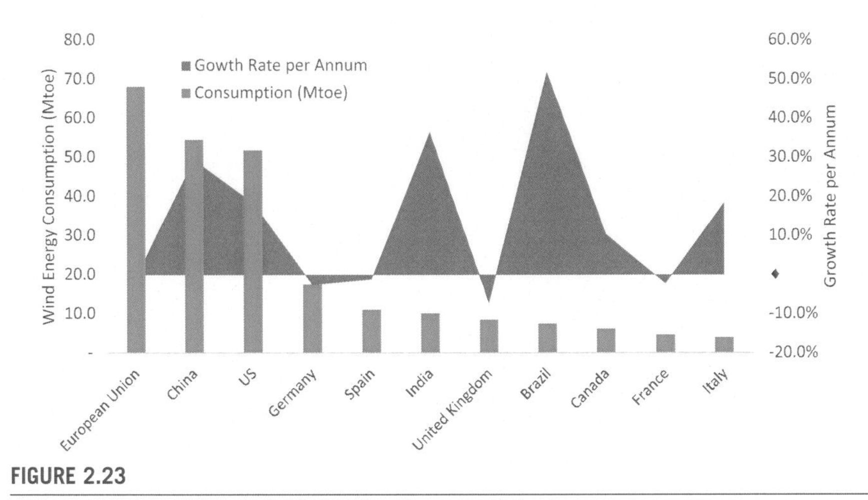

FIGURE 2.23

Wind energy consumption in 2016 for the top global consumers along with the growth rate per annum.

Data from: BP Statistical Review of World Energy. 2017. BP, London.

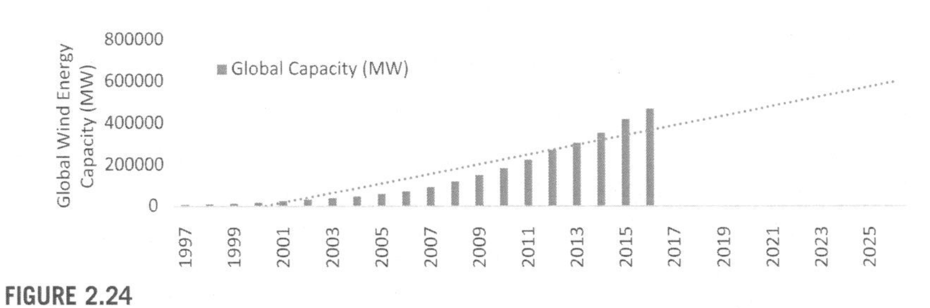

FIGURE 2.24

Global wind energy capacity along with projected capacity growth for the next decade.

Data from: BP Statistical Review of World Energy. 2017. BP, London.

the wind energy capacity is growing the world. Similar to the global wind energy consumption, the wind capacity is also rising continuously because of its viability and economic potential. Wind energy is the top renewable energy source after water in terms of global capacity.

2.4.2 Solar

Solar energy is another major renewable source, stemming from radiant light and heat from the sun, harnessed through various technological arrays. Evolving solar technologies include photovoltaics (PV), solar thermal energy, solar heating, molten solar power plants, and artificial photosynthesis. Solar energy is characterized as an important renewable energy source as some predict a solar revolution in the coming future following coal and oil. This source features two types of technologies: active

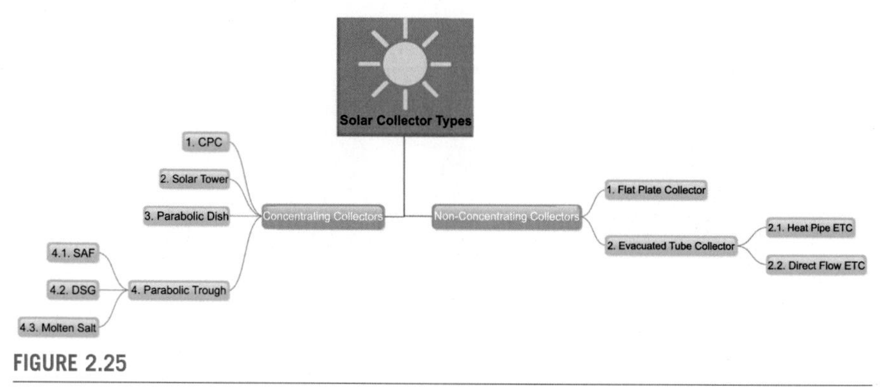

FIGURE 2.25

Classification of solar collectors and corresponding applications.

solar and passive solar methods. Active solar techniques include the use of photovoltaics, concentrated solar power, and water heating, whereas passive solar includes designing the building envelope to be solar-friendly or designing the building to naturally circulate air. Fig. 2.25 shows the various types of solar collectors and their respective uses and applications.

As mentioned earlier, the sun is earth's power bank and thus solar has started to play a vital role in offsetting electric grid especially during peak hours. However, because of its intermittent nature and the lack of reliable battery storage solutions, its role in electric generation is limited compared with nuclear, hydro, or fossil fuels. The effectiveness of solar energy depends on some factors including solar irradiance (W/m$^{2)}$, which is different from one location to another, size of the system, the direction the panel faces, the title of the panel, and the weather surrounding the system. For example, locations closer to the equator have higher solar irradiance than those closer the poles. Furthermore, cloud cover can affect the potential of the solar system by blocking the incoming irradiance. Currently, solar energy can be established on land and thus land availability is essential. However, floating solar systems are currently being developed in various regions of the world. Moreover, solar energy has become popular on already titled roofs. For example, homes with titled roofs that are facing the right direction have the potential to generate electricity by only investing in a solar system. Sometimes, governments subsidize and provide incentives for these types of initiatives. In addition, solar thermal energy can be utilized to provide various commodities such as water heating, HVAC, cooking, process heating, water treatment, and electricity generation. As for heating water, sunlight is harnessed from the sun and converted into heat. Water is heated up to 60°C and used for swimming pools and domestic hot water. Fig. 2.26 shows the historic global consumption of solar energy from 1989 until 2016 with projected consumption in the next decade.

Moreover, solar energy can be useful to offset the high-energy demand in heating, ventilation, and air conditioning of homes and buildings. In addition, solar can be useful in cooking, drying, pasteurization, heat producing through parabolic

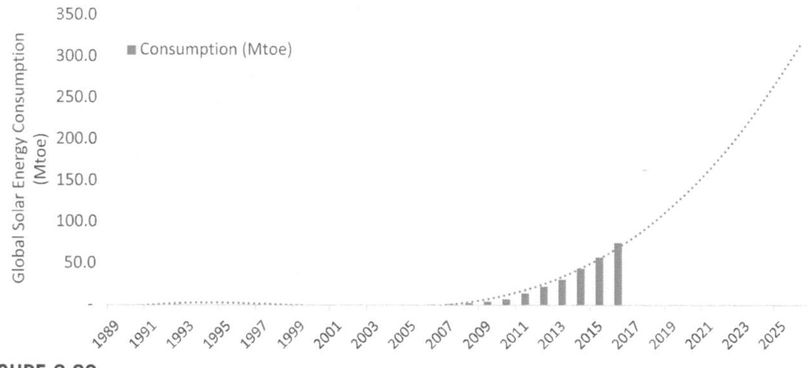

FIGURE 2.26

Global solar energy consumption between 1989 and 2016 with projected consumption for the next decade.

Data from: BP Statistical Review of World Energy. 2017. BP, London.

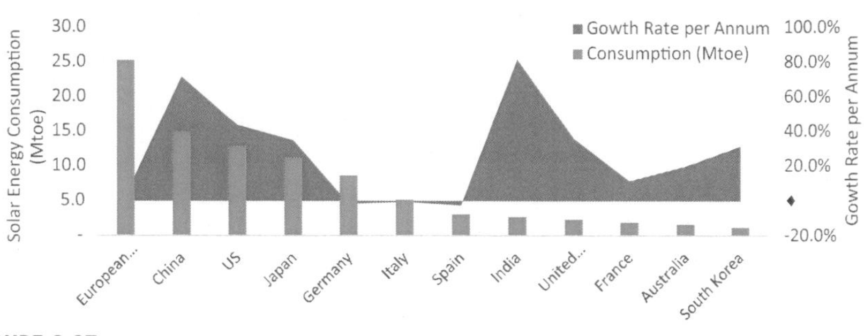

FIGURE 2.27

Solar energy consumption in 2016 for the top global consumers along with the growth rate per annum.

Data from: BP Statistical Review of World Energy. 2017. BP, London.

dishes, and electricity production. Solar rays are converted into electricity using photovoltaics or concentrated solar power (CSP). CSP systems are composed of a network of mirrors or lenses along with a tracking system to reflect the sunlight onto a small beam. Solar energy is also used in transportation, but further research is underway for better solar solutions. Furthermore, fuel production is another way to utilize solar. In fact, the solar energy is used to drive various chemical reactions. Fig. 2.27 shows the major consumers of solar energy in the world along with the rate of growth of solar energy in 2016 in these countries. It is evident that solar is rapidly growing in both India and China.

This is also another solution to store solar energy in various forms such as hydrogen, methanol, and ethanol. Solar energy storage is a significant area of research, and in fact, once this research matures, solar can become the leading energy source in the world. Because solar is an intermittent energy source, hybrid

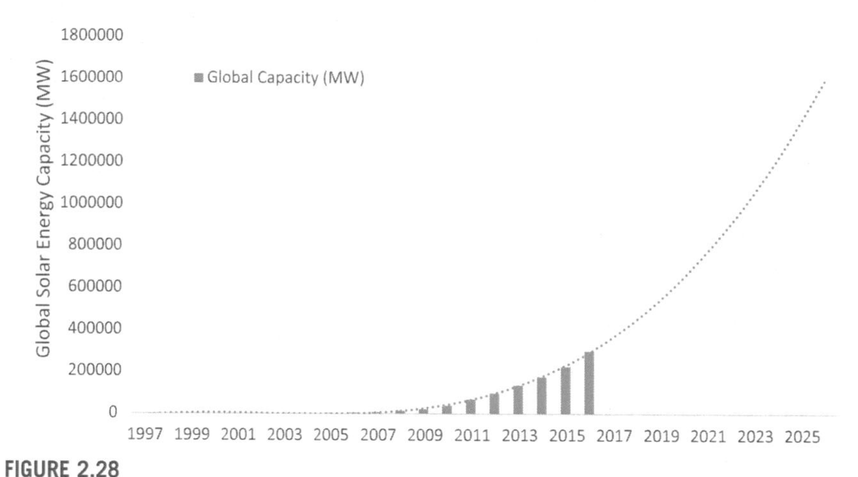

FIGURE 2.28

Global solar energy capacity along with projected capacity growth for the next decade.

Data from: BP Statistical Review of World Energy. 2017. BP, London.

systems have been designed to offset the intermittency. Hybrid systems combine the solar energy system such as PV or CSP to another energy source such as diesel, wind, or biogas. Hybrid systems are mostly found on islands to ensure energy reliability. All in all, solar energy is rapidly growing. In fact, Fig. 2.28 shows the world's solar capacity between 1997 and 2016. The rapid and steep growth is vivid indeed as it is projected to keep growing rapidly in the next decade.

2.4.3 Geothermal

Geothermal energy is simply the heat and power derived from the earth's interior. The thermal energy stored in earth's interior can be used for direct heating or cooling as well as electricity generation. Simply, the heat is extracted through a medium substance and used for space heating in the winter. On the other hand, the summer heat is extracted and discharged to the cooler ground. To have access to this energy source, deep wells must be drilled down until reaching the steam and hot water reservoirs. These are then used to drive turbines linked to electric generators. There are three types of geothermal power plants: steam, flash, and binary. Steam power plants extract steam directly from the ground and use it to power the turbines. Flash plants use the deep high-pressure hot water and pull it up toward a low-pressure water. The resulting steam is used to power the turbines. Finally, binary power plants use a secondary fluid with much lower boiling point than water. This causes the fluid to turn into vapor, which powers the turbine as a result. Geothermal plants energy normally come from liquid-dominated or vapor-dominated forms. The latter offers temperatures from 240 to 300°C, which enable the production of superheated steam. Fig. 2.29 shows the various types of geothermal plants along with highlights for each type.

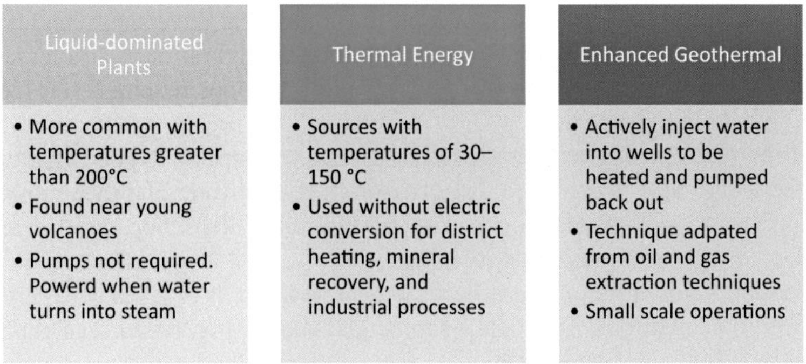

Liquid-dominated Plants
• More common with temperatures greater than 200°C
• Found near young volcanoes
• Pumps not required. Powerd when water turns into steam
Thermal Energy
• Sources with temperatures of 30–150 °C
• Used without electric conversion for district heating, mineral recovery, and industrial processes
Enhanced Geothermal
• Actively inject water into wells to be heated and pumped back out
• Technique adpated from oil and gas extraction techniques
• Small scale operations

FIGURE 2.29

Types of geothermal energy applications along with brief description for each.

Global Geothermal and Biomass Energy Consumption (Mtoe)
Consumption (Mtoe)
250.0
200.0
150.0
100.0
50.0
1965
1972
1979
1986
1993
2000
2007
2014
2021

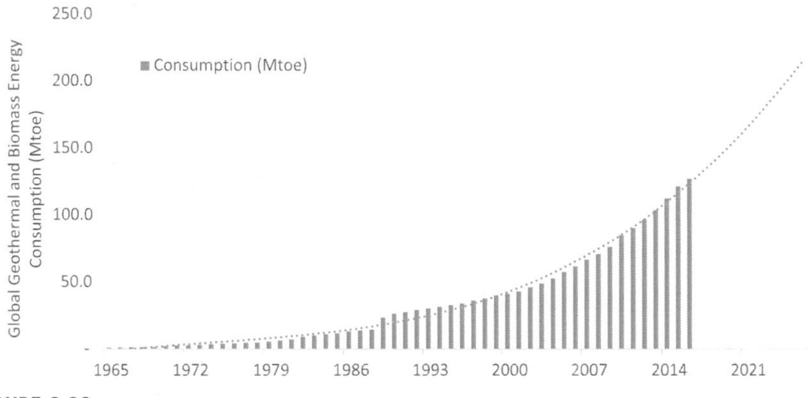

FIGURE 2.30

Global geothermal and biomass energy consumption between 1965 and 2016 with projected consumption for the next decade.

Data from: BP Statistical Review of World Energy. 2017. BP, London.

Geothermal is a reliable, cost-friendly, sustainable, and clean energy source; however, it is limited to areas near tectonic plate boundaries. Unlike solar and wind, geothermal is not an intermittent source and is regularly available. Global geothermal and biomass consumption is also growing with forecast of continuous growth in the next decade as observed in Fig. 2.30.

Concerns with geothermal are associated with the fluids used in these power plants such as the release of hydrogen sulfide and the disposal of some geothermal fluids, containing low levels of toxic materials. There are various ground loop configurations that can be used for geothermal systems including the following:

1. **Horizontal closed loop**: This is best used for rural areas with big landmass and moisture-rich soil that can be easily excavated. This configuration is not recommended for areas with dry sands.

2. **Vertical closed loop**: This is best used for homes that have small land area or those that do not have a better cost-effective loop configuration as an alternative. Vertical temperatures are more constant, and these loops require fewer feet piping than horizontal ones.
3. **Well-to-well/open loop**: This is used for homes with access to good quality well water, where heat is extracted directly from the well water and then returned into a second well. Temperatures extracted from the open loop system are higher than those from closed loops are.
4. **Lake/pond closed loop**: This is best for areas near ponds or lakes with very poor excavation conditions. The loop is submerged underwater, rather than buried in the ground.

Moreover, geothermal is very popular in China and Japan with growth rates ranging between 15% and 20% in 2016. However, its growth is depressing in India on the other hand. Fig. 2.31 shows the top consumers of geothermal and biomass energy in 2016 along with the rate of growth or depression for each country.

Although the European Union composes the world's top consumers of geothermal and biomass energy, the growth rate is negligible. On the other hand, China and Japan are considered modest consumers compared with the European Union, yet the growth rate for geothermal and biomass in these countries is the highest in the world. Geothermal and biomass capacity is slowly progressing globally. In fact, the capacity grew from 8 GW in 2000 to approximately 13 GW in 2018. Fig. 2.32 illustrates the global capacity growth for geothermal and biomass between 2000 and 2016 along with projected growth in the next decade.

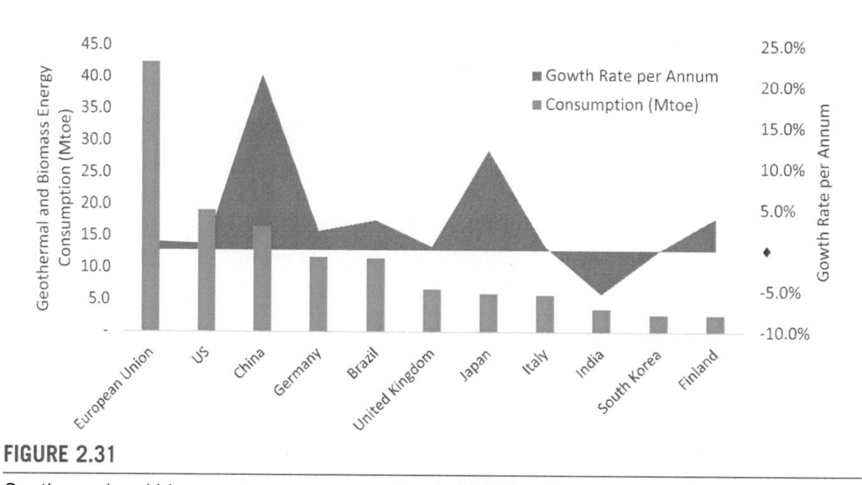

FIGURE 2.31

Geothermal and biomass energy consumption in 2016 for the top global consumers along with the growth rate per annum.

Data from: BP Statistical Review of World Energy. 2017. BP, London.

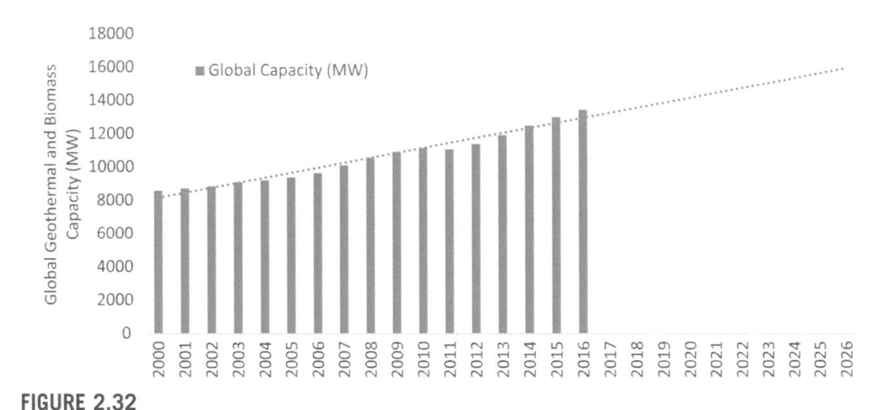

FIGURE 2.32

Global geothermal and biomass capacity along with projected capacity growth for the next decade.

Data from: BP Statistical Review of World Energy. 2017. BP, London.

2.4.4 Tidal and wave

Tidal and wave energy are similar because they both help generate electricity through the movement of ocean's waves and tides. Tide energy is driven by the gravitational pull of the sun and moon, which can be predicted well in advance. On the other hand, wave energy is the power harvested from the waves across the surface of the ocean. Because waves travel through the ocean, their arrival time at the wave power station can be more predictable than wind energy. Both types of energy are still being developed, and further research is needed for them to be commercially viable at large scale. Currently, they are very expensive, but the potential of wave and tidal power worldwide is enormous.

Wave energy resources are best between 30 and 60 degree latitude in both hemispheres, and the potential tends to be the greatest on western coasts. There are three types of wave energy technologies summarized in Fig. 2.33. Tidal and wave energy facilities produce zero GHG emissions unlike fossil-based fuels. Concerns are limited to marine ecosystems and fisheries. Furthermore, this energy source is being utilized for remote regions that are off-grid. Other socioeconomic benefits such as job creation, marine manufacturing, oceanography, and engineering are also expected from this energy source.

2.4.5 Biomass and biofuel

Biomass refers to the stored energy in organic material, and it is a renewable energy source. Biomass energy from wood or garbage can be directly burned and used for space heating of it can be converted into biofuels. These fuels can then be burned for energy. Crops on the other hand produce ethanol, which can be used in transportation. Fig. 2.34 illustrates various types of biomass and biofuels. Food crops are considered first-generation biofuels, as they are easily processed. Second-

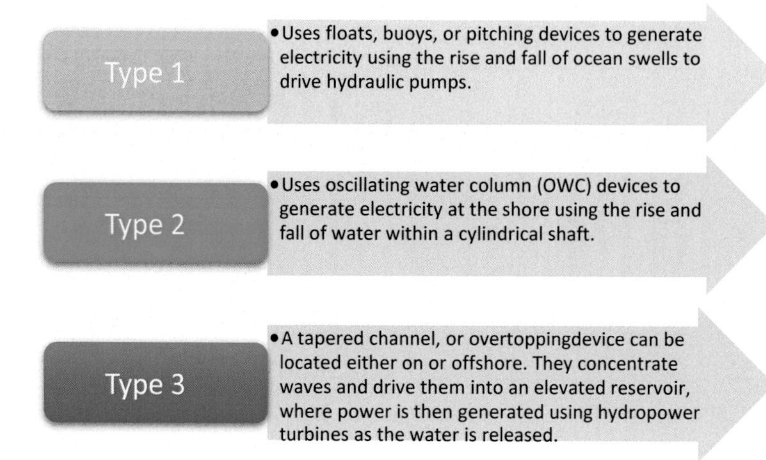

FIGURE 2.33

Types of wave energy technologies along with brief description for each.

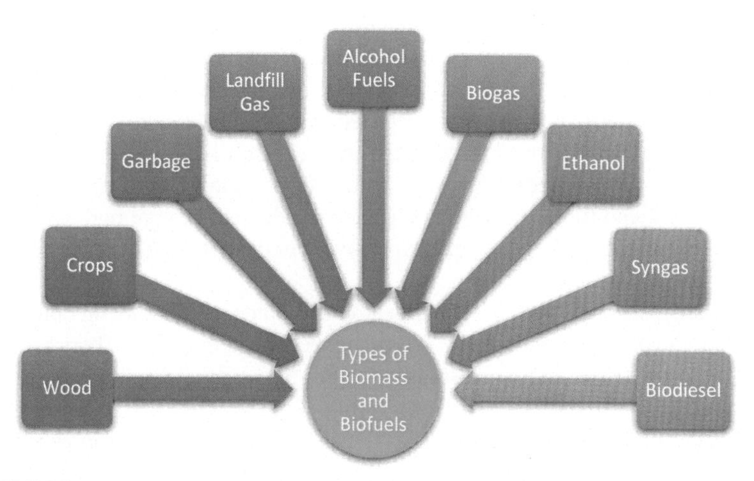

FIGURE 2.34

Types of various biomass and biofuels energy.

generation biofuels are produced from cellulosic materials such as wood and uneatable parts of plants. As these are hard to break down through fermentation, pretreatment is necessary.

Because of the burning of organic matter, this energy source emits GHGs, but it is classified as a renewable energy source because crops can be replaced with new growth. Furthermore, biomass has been the stepping stone for coal plants toward renewable energy. In fact, biofuels are the only viable replacement to fossil fuels, as they can be used in existing combustion engines. In addition, biofuels are combustible fuels created from biogas. The terms are usually used to describe liquid fuel such as ethanol or biodiesel, which are used as replacement for transportation fuels such as gasoline and diesel. Ethanol is an alcohol form by fermentation. It is used as a replacement for or an additive to gasoline. Biodiesel is an oil extracted by naturally occurring oils such as plants. It can be blended with diesel engines or added to diesel fuel. Aside from heat, there are other methods of converting biomass energy including chemical, biochemical, and electrochemical conversions. Fig. 2.35 demonstrates the various types of biomass conversions and example processes.

The global consumption of biofuels has also been undergoing a considerable growth from approximately 8000 ktoe in 1990 to approximately 82,000 ktoe in 2016. Fig. 2.36 shows the historic world's consumption of energy from biofuels and its projected growth for the next decade.

Unlike other sources of energy, biofuel is growing rapidly in third world countries. In fact, the growth rate of biofuels in Indonesia is the highest in the world in 2016, reaching 84%, whereas the growth in Argentina is 38%. Developed

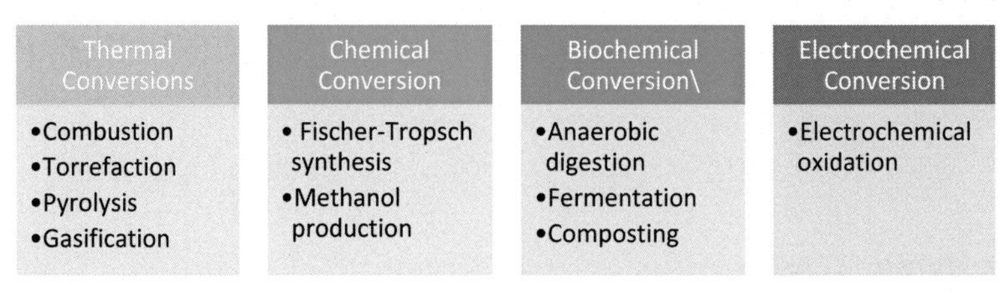

FIGURE 2.35

Types of various biomass conversions including different processes and methods.

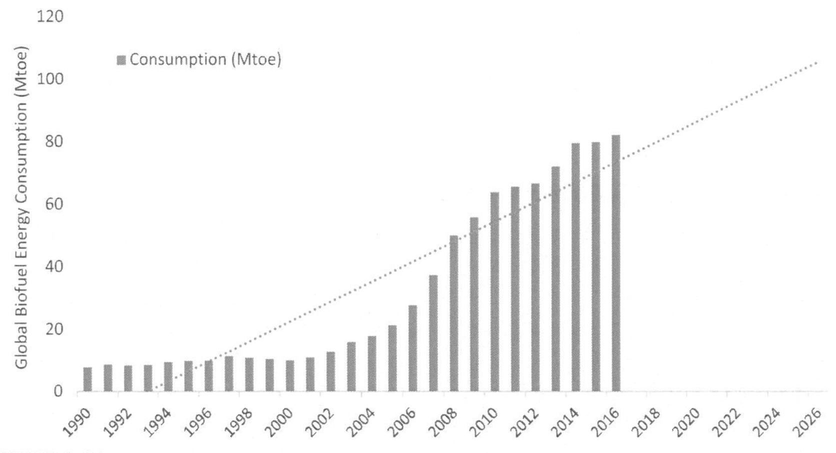

FIGURE 2.36

Global biofuel energy consumption between 1990 and 2016 with projected consumption for the next decade.

Data from: BP Statistical Review of World Energy. 2017. BP, London.

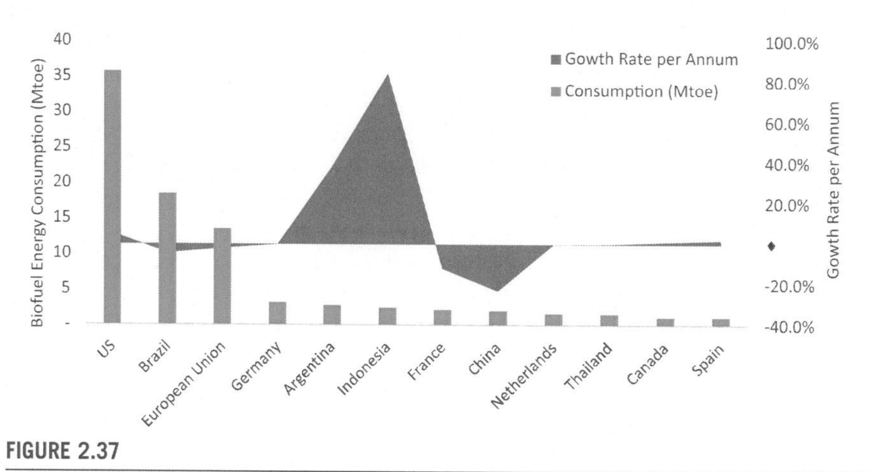

FIGURE 2.37

Biofuel energy consumption in 2016 for the top global consumers along with the growth rate per annum.

Data from: BP Statistical Review of World Energy. 2017. BP, London.

countries experience depression when it comes to biofuels, yet they remain the biggest consumers of energy using this source as illustrated in Fig. 2.37.

2.4.6 Hydro

Hydroelectricity is the energy that comes from hydropower, which takes advantage of gravity and the water cycle. Capturing energy through water has been practiced for thousands of years through watermills, for instance. Currently, hydropower is the main renewable energy source. In fact, it accounts for 70% of all renewable electricity and 16% of the world's total electricity. Moreover, hydropower is the only renewable energy source that can be utilized for baseload. Furthermore, hydropower is environmentally friendly, and GHG emissions are minimal. There are a number of generation methods including the following:

1. **Dams**: These are considered conventional methods. Electricity is generated as the water drives a water turbine and a generator. Water volume and height difference influences the generation.
2. **Pumped storage**: This method is used to supply electricity at high demand. Water is moved between reservoirs, and electricity is discharged when demand is highest.
3. **Run-of-the-river**: Small or no reservoir space. Electric generation is limited to water coming from upstream only. Excess water passes unutilized.

Dams create an opportunity for hydropower to be harvested. There are three main types of hydroelectric power plants. Impoundment facilities are large-scale facilities that utilize dams to store river waters. Upon release of water from the dams, waters power a turbine generator, which produces electricity. Fig. 2.38

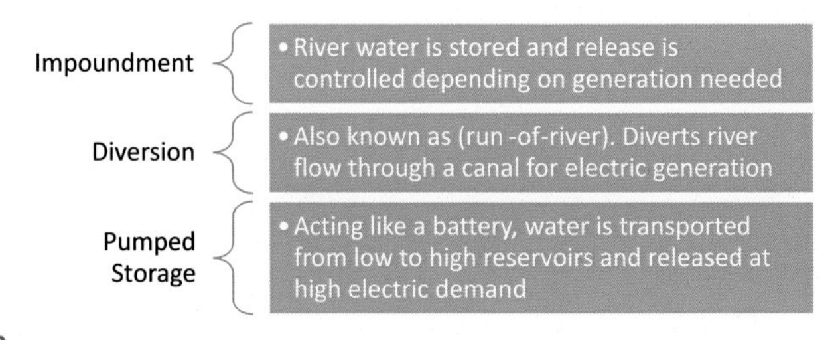

FIGURE 2.38

Types of hydroelectric power plants depending on generation method.

shows the main three types of hydroelectric power plants. The diversion type essentially draws water from a river by diverting its path through a canal. Water at high speeds and potential energy powers a generator, which generates electricity and then the water exits through an output channel. The third type involves transporting water from lower reservoirs to upper reservoirs, where they gain potential energy. The water remains stored in the upper reservoirs and are released downhill to power a generator during peak power demands. Fig. 2.38 shows the types of hydro power plants depending on the method of generation used by each power plant.

The amount of available power is a function of the hydraulic head and the rate of fluid flow. Therefore, the power of falling water can be calculated using the following equation:

$$P = \eta \rho Q g h \tag{2.2}$$

where P is power in watts, η is the dimensionless turbine efficiency, ρ is the water density in kg/m^3, Q is the flow in m^3/s, g is the gravitational acceleration of the water, and h is the height difference between the input and output outlets. The global hydro consumption totaled 910 Mtoe in 2016 (BP Statistical Review, 2017). Furthermore, this consumption steadily grew, illustrating a positive linear relationship as illustrated in Fig. 2.39.

Furthermore, the biggest global consumer for hydro energy in 2016 is China followed by Canada and Brazil, whereas Norway and the United States were the countries with the highest growth rates per annum. Fig. 2.40 shows the top consumers of hydro energy in the world in 2016.

2.4.7 Hydrogen

Hydrogen is the most abundant element in the universe. Simply with a single proton and another single electron, hydrogen is plentiful, yet does not occur naturally as a gas on the earth. It is always coupled with other elements such as oxygen, producing water, for example. Furthermore, hydrogen is found in various organic matter such as hydrocarbons. As hydrogen is a zero-emission fuel, it can be used in internal

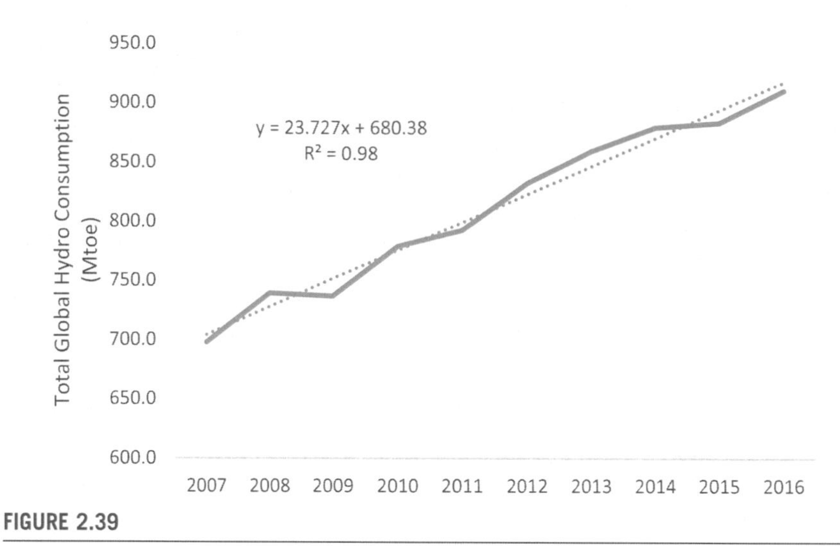

FIGURE 2.39

Total global hydro consumption between 2007 and 2016.

Data from: BP Statistical Review of World Energy. 2017. BP, London.

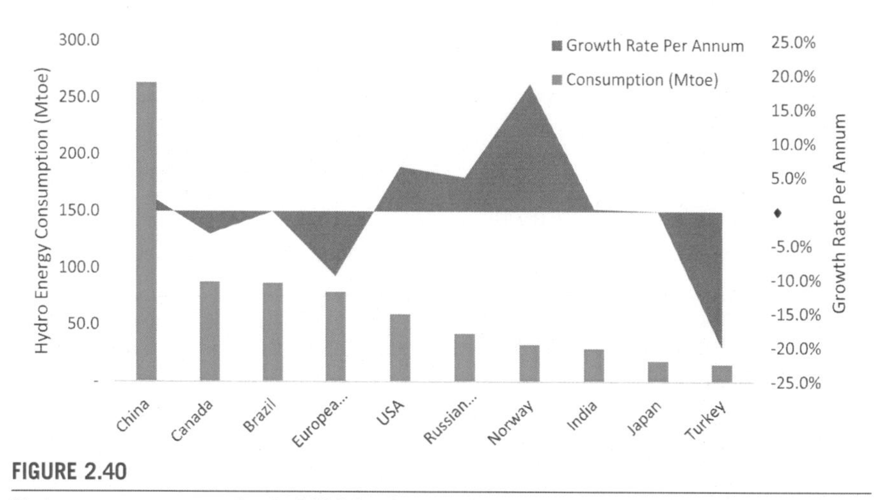

FIGURE 2.40

Hydro energy consumption in 2016 for the top global consumers along with the growth rate per annum.

Data from: BP Statistical Review of World Energy. 2017. BP, London.

combustion engines or electrochemical cells, and in fact, it is currently being used in some commercial fuel cell vehicles. Hydrogen separation or production can be conducted through various means.

In fact, hydrogen production through renewable means has become more popular; however, the majority of hydrogen production comes from utilizing

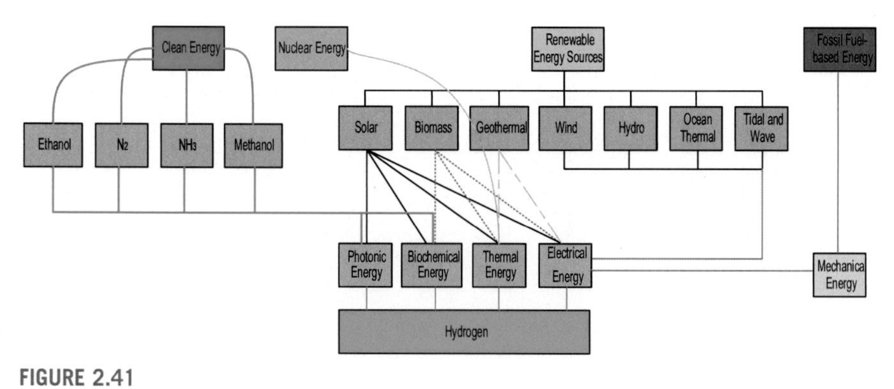

FIGURE 2.41

Hydrogen production pathways using various energy sources.

Modified from Dincer and Acar, 2015

hydrocarbons via thermochemical processes. Solar energy can be used through photochemical processed to produce hydrogen. Furthermore, biomass can be utilized through biochemical energy for hydrogen production using dark fermentation, thermochemical conversion, anaerobic digestion, photofermentation, and sequential dark and photofermentation processes. Moreover, geothermal energy can also be used via thermochemical water splitting or electrolysis. In summary, hydrogen production needs primary energy sources to be utilized through specific processes to produce useable quantity of hydrogen, which can be later used of stored. Fig. 2.41 shows the various pathways for hydrogen production using renewable energy sources. All in all, the world is moving toward renewable energy. In fact, the world's consumption of other renewable sources including hydrogen totaled 419 Mtoe in 2016 (BP Statistical Review, 2017). Fig. 2.42 shows the top countries globally that consumed the most in terms of other renewable sources in 2016.

2.5 Closing remarks

In conclusion, energy sources vary in abundance as they can also be configured in various integrated ways. In fact, multiple sources can be utilized to provide a common service or commodity. Furthermore, fossil-based fuels remain to be important sources of energy in the world as renewable energy sources are also gaining more acceptance. Environmental impacts of fossil fuels make them unfavorable and put them at a disadvantage when compared with renewable energy sources, which are also abundant, clean, and available. Moreover, unconventional sources of energy such as hydrogen can be produced through various ways including existing

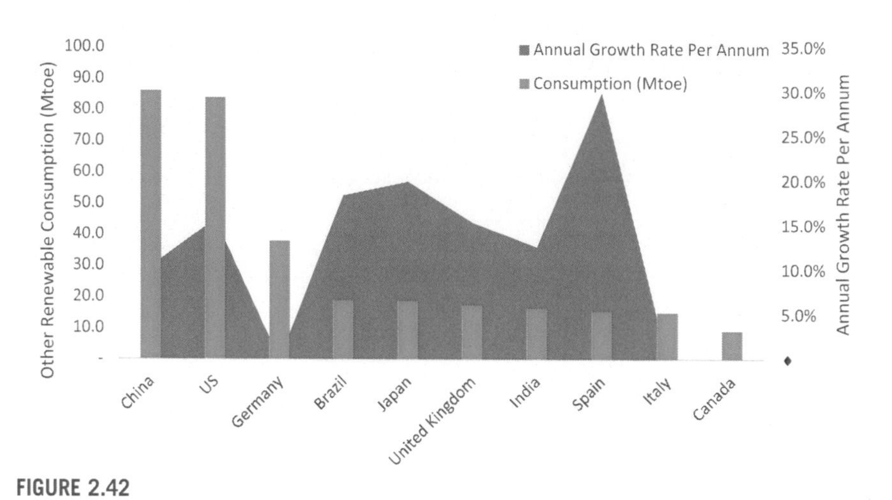

FIGURE 2.42

Other renewable energy consumption in 2016 for the top global consumers along with the growth rate per annum.

Data from: BP Statistical Review of World Energy. 2017. BP, London.

renewable and nonrenewable sources. In fact, the growth rate per annum for solar energy in 2016 approximates 30% followed by wind energy at 15%, whereas the rest of energy sources are much below 5%. This is a clear indicator that renewable energy and specifically solar is gaining more social acceptance and evolving to be a viable energy solution for various processes.

CHAPTER

Energy systems

3

3.1 Introduction

The primary function for an energy system is to provide energy services to end users. This includes the main commodities required for a society such as electricity, heating, cooling, water treatment, and energy storage. Furthermore, these commodities are supplied to five main sectors, including residential, industrial, commercial, agricultural, and transportation sectors, which make up the primary components for modern societies. Moreover, energy systems can be classified as systems that supply electricity, heating, or cooling applications. Fig. 3.1 shows a comprehensive overview of energy systems.

3.2 Power-generating systems

Electromechanical generators driven by heat engines at power plants are the conventional method to produce electricity. These heat engines can be fueled by different means such as combustion, nuclear fission, or kinetic energy from water for instance. Other fuel sources include geothermal power and solar photovoltaics. Generated electricity uses transmission lines and poles to deliver this service to a wider

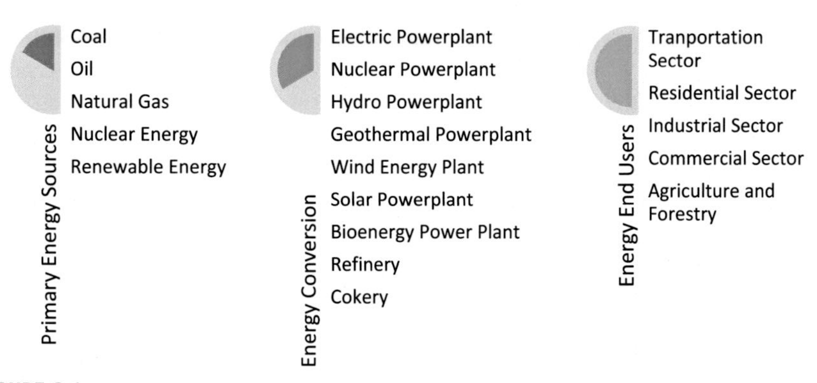

FIGURE 3.1

Comprehensive overview of energy systems from the primary sources to the end users.

Energy Sustainability. https://doi.org/10.1016/B978-0-12-819556-7.00003-6

circumference. Thermal power stations allow for mechanical power to be produced from the transformation of thermal energy using a heat engine. These stations can be classified in different ways. For instance, they can be classified by heat source. Under this classification, there are solar thermal stations where sunlight is used to boil the water into very high temperatures and turn it into steam, which is then used to turn the generator. Furthermore, waste heat is also concentrated occasionally and used in a steam boiler and turbine for power generation. Moreover, biomass-fueled power plants may be fueled by municipal solid waste, methane from landfills or other biomass forms. In addition, geothermal plants use steam extracted from hot underground layers to power the generators. On the other hand, nuclear power plants are more complicated as they use the heat generated from the fission processes to create steam, which is used to operate a steam turbine and generator. Other than these sources, the conventional method for power generation is the fossil-fuel power generation. In this method, fossil-based fuels such as coal, oil, or natural gas undergo combustion to power a steam turbine and generator.

Another way to classify power-generating plants is by the prime mover. Steam turbine plants produce approximately 90% of all electric power in the world. On the other hand, gas turbine plants use dynamic pressure resulting from gas flow to directly operate the turbine. Combined cycle plants have both options aforementioned combined. In this case, the overall efficiency of the plant is enhanced. Isolated communities use internal combustion reciprocating engines to supply power. Another way to classify power generation plants is by duty. For instance, some plants are designed for base load and thus, they run continually. Other plants are designed for peaking times and thus, they only run for some hours during the day to fulfill the daily electricity demand. The last type of plants is the load following power plant, which follows the variations of the daily or weekly load. These plants are more flexible than base load plants and can be more cost-effective than peaking power plants. Therefore, power generation can be achieved through various means and by using different sources. Fig. 3.2 illustrates the world's power generation in 2016 by source.

Coal remains a leading source for power generation especially in Asia, whereas natural gas is the second popular source for electric generation. All in all, fossil-based fuels including coal, natural gas, and oil account for 65% of the total global power generation in 2016. On the other hand, renewables including hydro account for almost a quarter of the world's overall power generation. Besides, nuclear fission is also an important source for power generation as it accounts for 10% of the total power generation. The total power generation for this year is recorded as 24.973 TWh (IEA, 2018).

3.2.1 Fossil-fuel power plants

In fossil-fuel power stations, chemical energy stored in natural gas, oil, and coal is converted to thermal energy by burning and consequently to mechanical energy through a turbine, which powers a generator. After that, electric energy is produced and supplied to various sectors. This type of power stations dominates the global

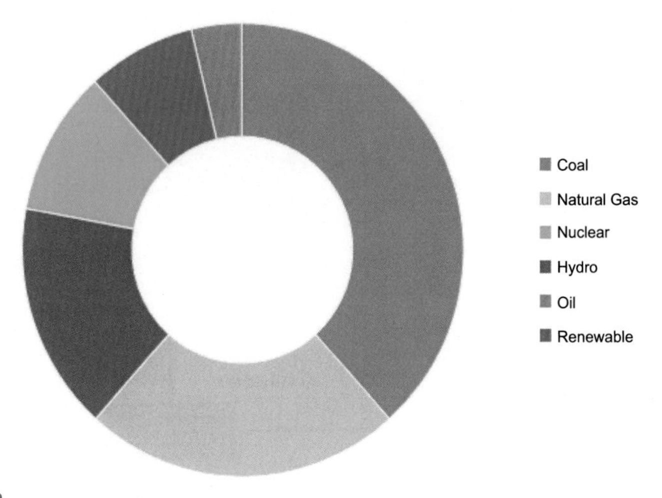

FIGURE 3.2

World's power generation by source in 2016.

Modified from IEA, 2018.

market as fossil fuel power stations outnumber nuclear, geothermal, solar, and biomass power plants. Moreover, coal is the most abundant fossil fuel in the world, making it a leading source for power generation. It is in fact cheap and can be found in abundance in regions such as China, India, and the United States. On the other hand, natural gas is abundant in regions such as Russia, Qatar, and Iran. Although coal is readily useful for power generation, natural gas is handy for transportation fuel and water and space heating. Coal can be gasified or converted to liquid to provide space and water heating, but it cannot replace natural gas and the process of gasification or liquefying poses higher costs and inefficiencies. Fuel can be transported through various means including trucks, rail, or through pipelines. Once at the power plant, fuel is processed before entering the power generation cycle. For example, coal is crushed into small pieces. Coal preheating or pulverization can also take place before the cycle. Rankine cycle is the ideal thermodynamic cycle for vapor power plants that converts heat into mechanical work. Both electricity is produced from the turbine as well as heat is rejected from the condenser as illustrated in Fig. 3.3.

The Rankine cycle is a model that is used to forecast the steam turbine's performance. In this model, the process, which the steam-operated heat engines undergo is described. Heat sources for this cycle as mentioned previously include fossil-based fuels or other fuels such as nuclear fission. On the other hand the cooling sources for this cycle include cooling towers and large bodies of water such as a sea or a river. The Rankine cycle features four different states as follows (Çengel et al., 2019):

1-2 Isentropic compression in a pump
2-3 Constant pressure heat addition in boiler

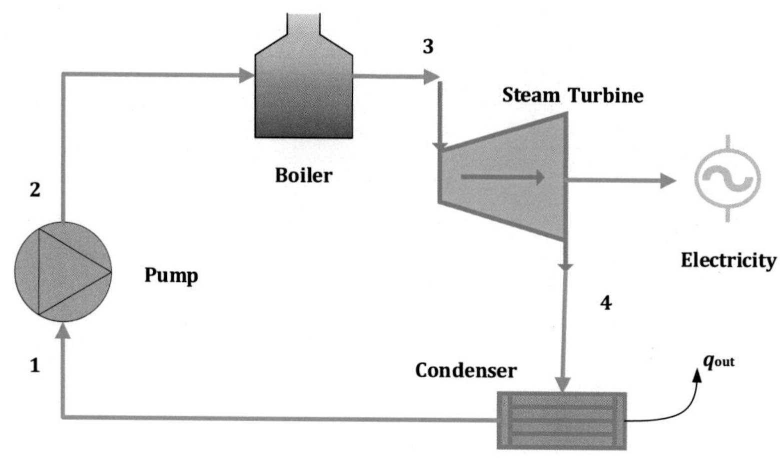

FIGURE 3.3

Ideal Rankine cycle with all components and state points.

3-4 Isentropic expansion in a turbine
4-1 Constant pressure heat rejection in a condenser

In this cycle, water is fed through piper toward the boiler. In conventional steam-electric plants, the condenser removes the latent heat of vaporization from the incoming steam and uses that enthalpy to preheat the feedwater to reduce irreversibilities involved in the cycle. This allows for improving the thermodynamic efficiency of the system. Once the water is inside the boiler, it goes under extreme heat and pressure, converting the water to steam and adding the latent heat of vaporization. The water in the boiler passes through the economizer, which is a section in the convection pass. Then, it goes to the steam drum and eventually turns into steam through the water walls. Steam separators block any water droplets from entering the turbine as that causes damage and erosion. Only steam goes through to the turbine. The steam powers a high-pressure turbine followed by a low-pressure turbine, which generates electricity. As the condenser also ejects heat, combined heat and power (CHP) systems are introduced. These systems are considered cogeneration systems because they can produce two distinct useful commodities from the same source. These systems normally provide electric power as well as heating for water, domestic heating, or industrial heating. Once again, this cogeneration system is considered cost-effective, efficient compared to a conventional single generation system, and consequently more environmentally friendly.

Although this type of power plants is considered useful and practical, it poses an environmental threat as they cause the emission of various toxic gases and particulate matter such as SOx, NOx, CO_2, CO, and other organic gases and hydrocarbons. Coal combustion contributes directly to the introduction of acid rain and air pollution. Furthermore, although natural gas is much less harmful than coal due to its chemical composition, it also accounts for significant amounts of greenhouse gas emissions. This critical setback of this type of power plants introduced the idea of greening fossil fuel power plants. Many countries that signed the Kyoto Protocol and the Montreal Protocol are rapidly converting their coal power plants to biomass or waste power plants. Some countries like Canada are shutting down coal power plants and removing this fuel source completely. On the other hand, natural gas power plants are converted to biogas. The ultimate objective behind these conversions is to mitigate pollution caused by these plants as well as enhance their efficiencies and useful output.

3.2.2 Nuclear power plants

Nuclear power stations are another type of thermal power stations, where heat is used to power a steam turbine that converts the mechanical energy to electricity. These types of power stations are considered base load power stations, as they cannot be turned off rapidly or easily. Rather, they supply the base load and are slow to be dispatched. Nuclear power plants are not as widespread across the world due to various reasons. According to IEA (2018), there are 450 nuclear power stations across 31 countries in the world. In this system, nuclear fission heats the reactor coolant, which then heats the feedwater. The heated water turns into pressurized steam and powers a multistage turbine that generates electricity. After this step, the remaining vapor is condensed at the condenser, which also acts as a heat exchanger and is usually connected to a secondary side such as a lake or a sea. The water is then pumped back to be heated as the cycle continues.

The most important component in the nuclear power plant is the nuclear reactor. In this component, the chain reaction that produces heat is usually triggered by Uranium. Uranium occurs in two isotopic forms, either U-238, or U-235, which differ in the number of protons and behavior. Furthermore, nuclear fission is the process when larger fissile atomic nucleus absorbs neutrons, causing nuclear fission.

This means that the heavy nucleus is split into two lighter nuclei, which also cause the emission of gamma radiation, kinetic energy, and free neutrons, which contribute to further chain reactions. Nuclear reactors can be classified in different means. The main classification is based on the reactor type. Table 3.1 outlines the various reactor types in detail.

Nuclear reactors are composed of three main components, namely the control rods, fuel elements, and moderator. As nuclear reactors vary in type, they also have distinct number of control rods and different types of fuel elements.

As illustrated in Fig. 3.4, a pressurized water reactor features a pressurizer and a specialized pressurized vessel. This type of reactor uses light water as the coolant and is mainly used for electricity generation. Furthermore, it is also capable of producing nuclear submarines and naval vessels. Moreover, this type of nuclear reactor is the most widely used worldwide.

A subsidiary type of the pressurized water reactor is the pressurized heavy water reactor illustrated in Fig. 3.5. These reactor types use pressurized isolated heat transport loop and heavy water as the moderator. Natural uranium oxide is used as the

Table 3.1 Various reactor types and their associated properties and specifications.

Reactor type	Moderator	Steam cycle efficiency	Fuel	Coolant		
				Outlet temperature [°C]	Heat extraction	Pressure [psi]
Pressurized water reactor (PWR)	Water	32%	Enriched UO_2	317	Water	2235
Boiling water reactor (BWR)	Water	32%	Enriched UO_2	286	Water	1050
Pressurized heavy water reactor (PHWR)	Heavy water	—	Natural UO_2	310	Heavy water	1450
Gas-cooled reactor (AGR)	Graphite	42%	Natural U (metal), enriched UO_2	650	CO_2	600
Gas-cooled reactor (magnox)	Graphite	31%	Natural U (metal), enriched UO_2	360	CO_2	300
CANDU	Heavy water	30%	Unenriched UO_2	305	Heavy water	1285
Light water graphite reactor (RBMK)	Graphite	31%	Enriched UO_2	284	Water	1000
Fast neutron reactor (FBR)	—	—	PuO_2 and UO_2	510	Liquid sodium	—

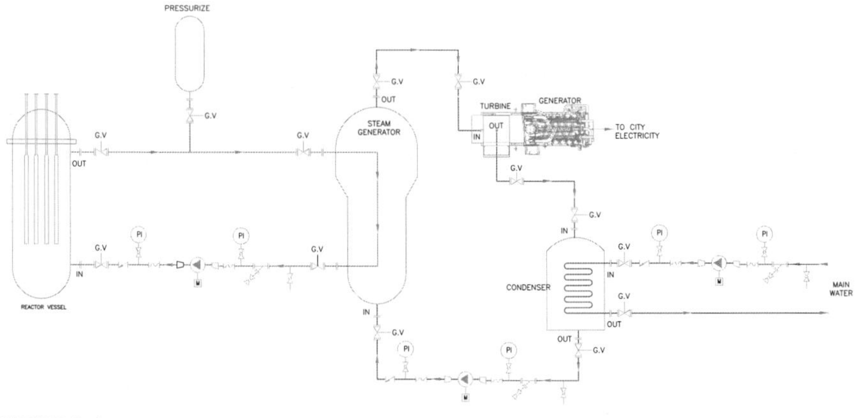

FIGURE 3.4

Schematic of a pressurized water nuclear reactor including its main components.

fuel. There are 49 nuclear reactors of this type worldwide (World Nuclear Association, 2018). This reactor type features the use of heavy water that enters the reactor vessel and then exit to power a steam generator. The generated steam powers a turbine that consequently generates electricity as a useful output.

Heat emission by nuclear fission is used to boil water in the boiling water nuclear reactor. Boiled water consequently transforms into steam, which generates a turbine as illustrated in Fig. 3.6.

There are 14 gas-cooled nuclear reactors worldwide. This type of nuclear reactor uses carbon dioxide as coolant and graphite as a moderator and natural uranium as the fuel. Carbon dioxide circulates through the system and reaches a very high temperature of 650°C. Fig. 3.7 shows the schematic of the advanced gas-cooled reactor (AGR).

Adopted from the magnox reactor, the AGR is a second-generation gas-cooled reactor that also used graphite as the moderator and carbon dioxide as the coolant. Fig. 3.8 illustrates a schematic of the advanced gas-cooled reactor.

The last type of nuclear reactor covered in this book is the light water graphite-moderated reactor. This type of reactor was developed by the Soviets and uses water as the coolant, graphite as the moderator, and enriched uranium oxide as the fuel. Fig. 3.9 provides a schematic of this type of reactor and its main components.

The most common type of reactor is the pressurized water reactor (PWR), which make up 63% of all reactors in the world (IAEA, 2017). Fig. 3.10 shows the number of reactors for each type as of 2014. It is clear that PWR dominates the types of nuclear reactors in the world. This makes it imperative that this type be studied in detail.

In the previous chapters, we presented a different classification for nuclear reactions, which is the classification by generations. Nuclear reactors can be classified as generation I, II, III, and IV reactors depending on the age of the reactor. Unfortunately, this energy source also brings environmental concerns as experienced

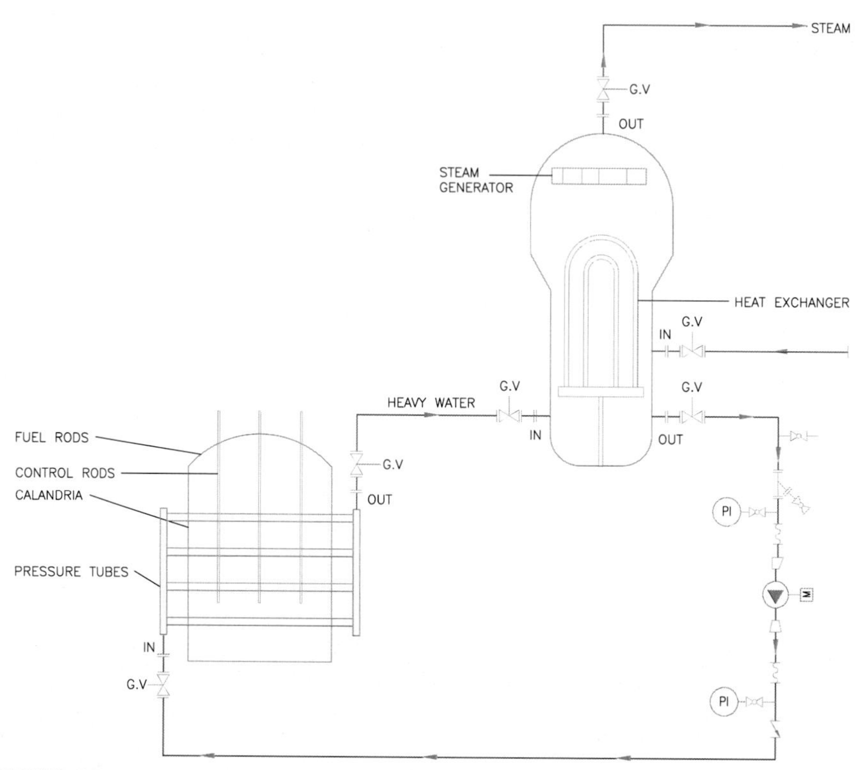

FIGURE 3.5

Schematic of a pressurized heavy water nuclear reactor including its main components.

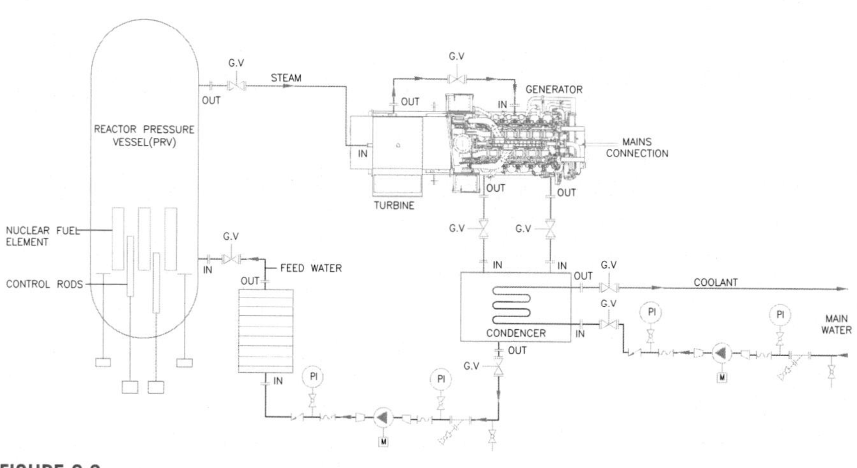

FIGURE 3.6

Schematic of a boiling water nuclear reactor including its main components.

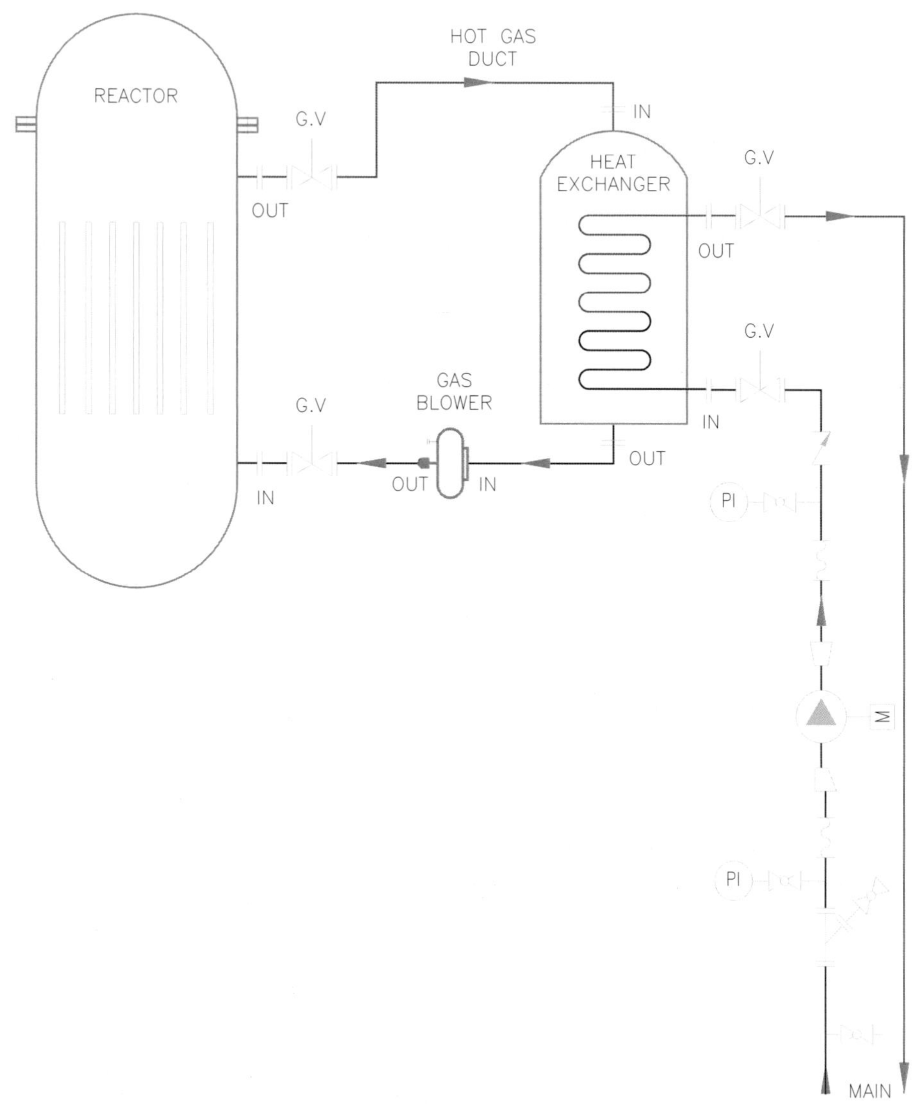

FIGURE 3.7

Schematic of a gas-cooled nuclear reactor including its main components.

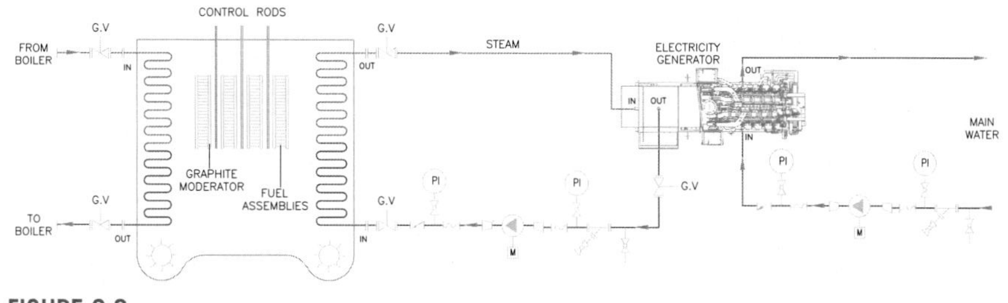

FIGURE 3.8

Schematic of an advanced gas-cooled nuclear reactor including its main components.

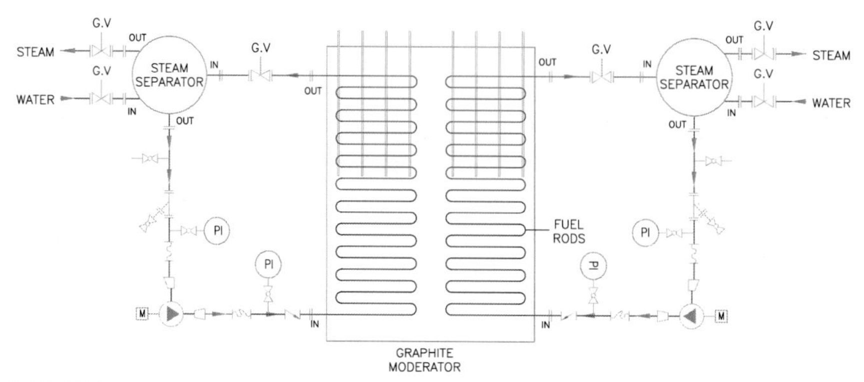

FIGURE 3.9

Schematic of a light water graphite nuclear reactor including its main components.

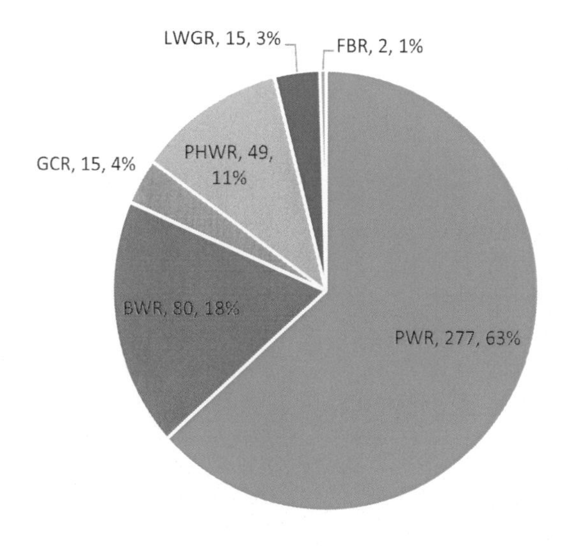

FIGURE 3.10

Number of reactors per type in the world in 2014.

Data from IAEA Annual Report, 2017. Vienna: International Atomic Energy Agency.

previously in the Chernobyl disaster in 1986. Therefore, some countries may be scaling down their nuclear power plants and decommissioning them, while increasing their renewable power plants instead. However, despite the challenges that nuclear power plants bring, they remain a competent and a leading option for power generation, especially for base load generation.

3.2.3 Geothermal power plants

Geothermal power plants rely on underground heat, which is stored in rocks and groundwater to power steam turbines, which subsequently generate electricity.

These power plants function very similar to fossil fuel and nuclear power plants; however, in this case the heat source is different. Essentially, hot water or steam is extracted from the earth through a series of wells and is used to power the steam turbine by allowing hot, pressurized geothermal fluid to expand rapidly and provide rotational energy to turn the blades of the turbine on the shaft. The mechanical energy caused by the rotation of the blades on the shaft is used to spin magnets inside a large coil and create electric current. Therefore, the turbine and the generator are crucial components of the geothermal system, which transform geothermal energy to electric energy. Water is typically reinjected back through a reinjection well. There are three different types of geothermal plants: dry steam plants, flash steam plants, and binary cycle power plants. Fig. 3.11 shows the various types of geothermal plants and their specifications.

Flash steam plants can be further designed as single, double, triple, quadruple, and quintuple system, each with and without reinjection. Single flash geothermal power system uses hot geothermal water at various temperatures not less than 180°C as the heat source. It is assumed that steady operating conditions exist while kinetic and potential energies are negligible. Fig. 3.12 illustrates the system with the various state points.

Hot water is extracted from a geothermal well and passes through a flash chamber and a separator before activating a steam turbine. Once steam exits, it undergoes condensation and is reinjected back underground. After observing the single flash system with and without reinjection, we are now investigating a double flash system with reinjection phenomenon. Fig. 3.13 illustrates the system and its specifications.

As can be observed, there are two separators in this system and two inlet streams to the turbine while the reinjection well remains the same as in the single flash system. Furthermore, triple flash steam power system with reinjection can be observed in Fig. 3.14, with the different state points and components for this system.

Dry Steam Plants

- Use dry steam, that is naturally produced in the ground
- Oldest types of geothermal plants
- Can only be used where underground temperatures are extremely high

Flash Steam Plants

- Most common due to the lack of naturally occuring high-quality steam
- Water must be over 180 degrees Celsius
- Higher cost associated with construction and maintenance

Binary Steam Plants

- Expected to dominate geothermal plants
- Ability to utilize lower water temperatures
- Use a secondary loop, containing a fluid with lower boiling point

FIGURE 3.11

Types of geothermal power plants including main features for each type.

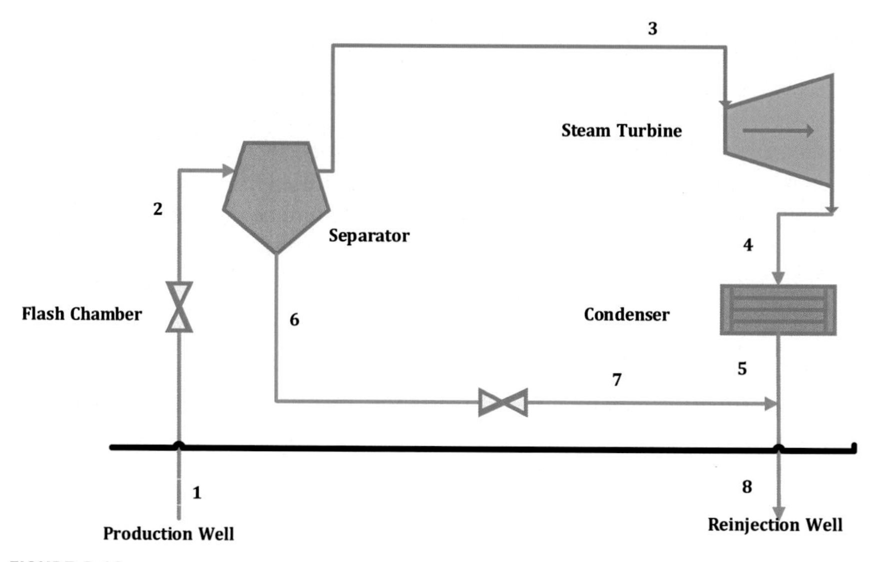

FIGURE 3.12

Single flash steam power system with reinjection including all state points.

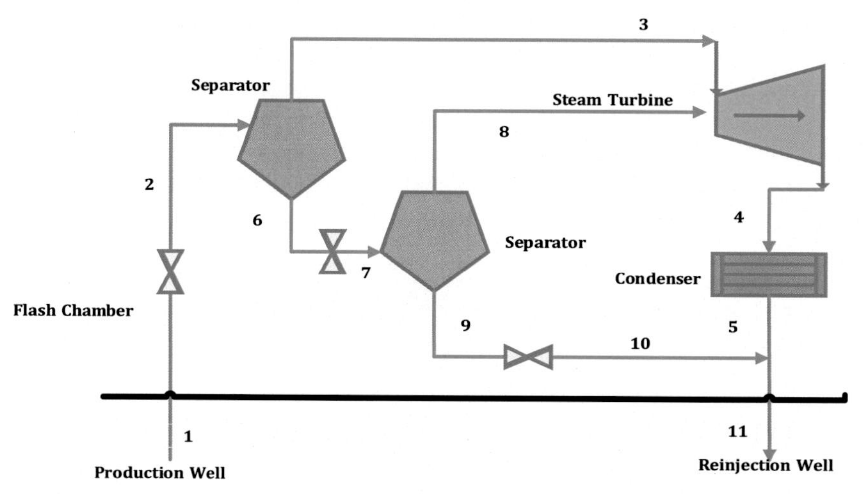

FIGURE 3.13

Double flash steam power system with reinjection including all state points.

The following is an illustration of a quadruple flash steam power system with reinjection. Fig. 3.15 shows the system along with all the state points including four separators. It is also important to note that the generator indicated in these figures actually reflects a number of generators that is connected with each individual separator. The steam coming from each separator varies in thermodynamic properties.

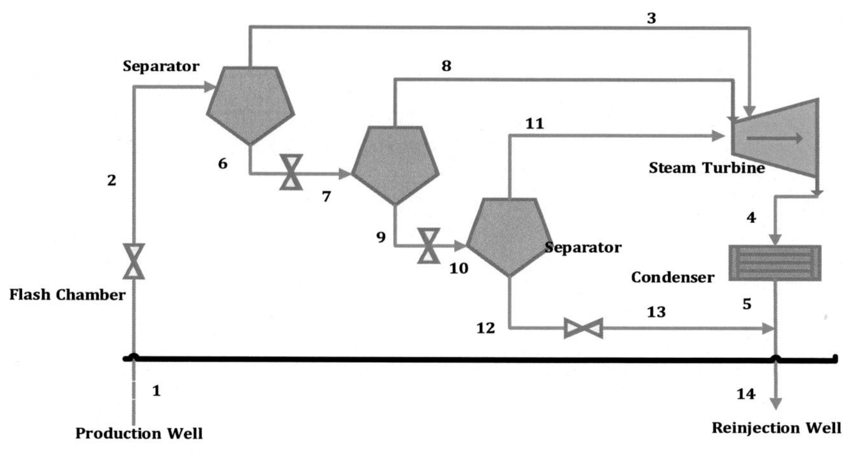

FIGURE 3.14
Triple flash steam power system with reinjection including all state points.

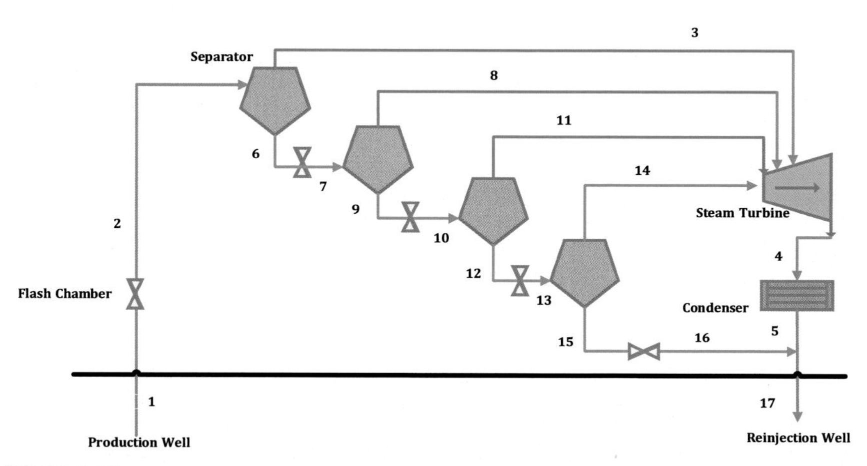

FIGURE 3.15
Quadruple flash steam power system with reinjection including all state points.

The following illustration proposes a quintuple flash steam power system with reinjection option. Fig. 3.16 shows the depicted system with all state points.

The dry steam plant is more simplified than this model as the steam is naturally sufficient to power the turbine. Such steam intensity can only be found in specific sites along the oceanic ridges. Fig. 3.17 illustrates the different components of the dry steam power plant.

On the other hand, binary geothermal plants utilize two fluids, one with a lower boiling point. This configuration allows for the utilization of lower well temperatures. Fig. 3.18 illustrates the main components of the binary geothermal system.

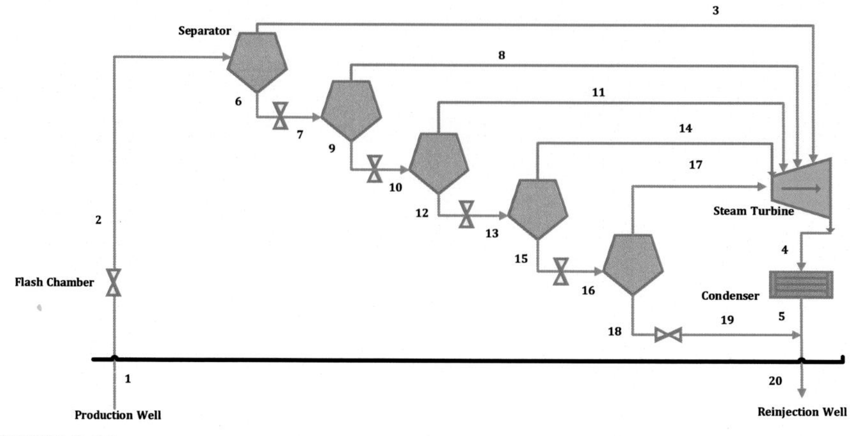

FIGURE 3.16

Quintuple flash steam power system with reinjection including all state points.

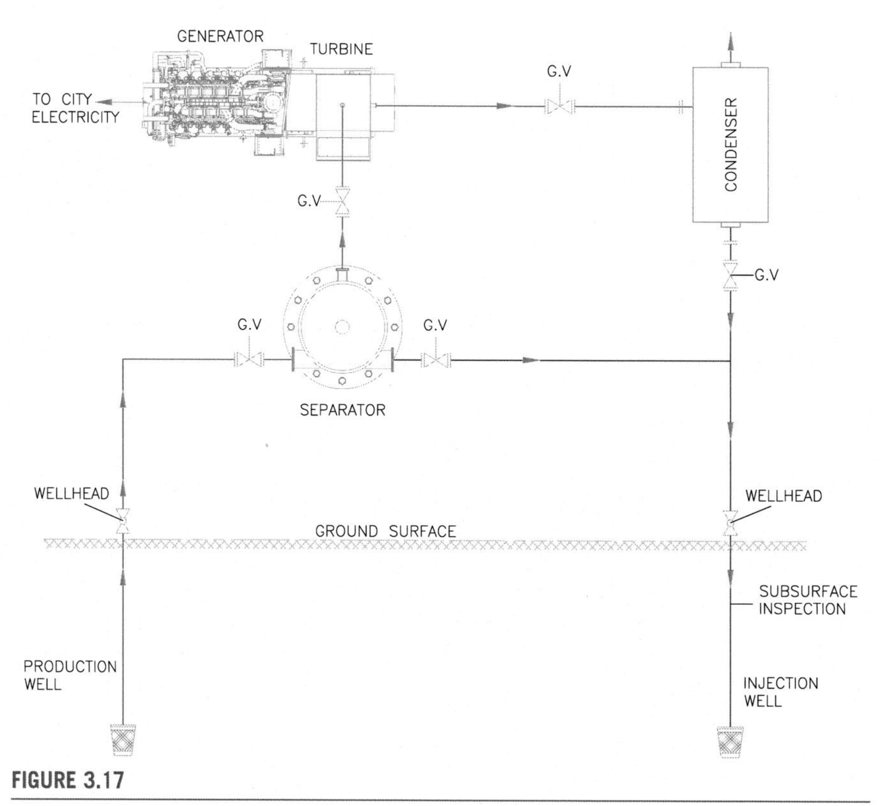

FIGURE 3.17

Dry steam geothermal power plant indicating the stages of steam and water.

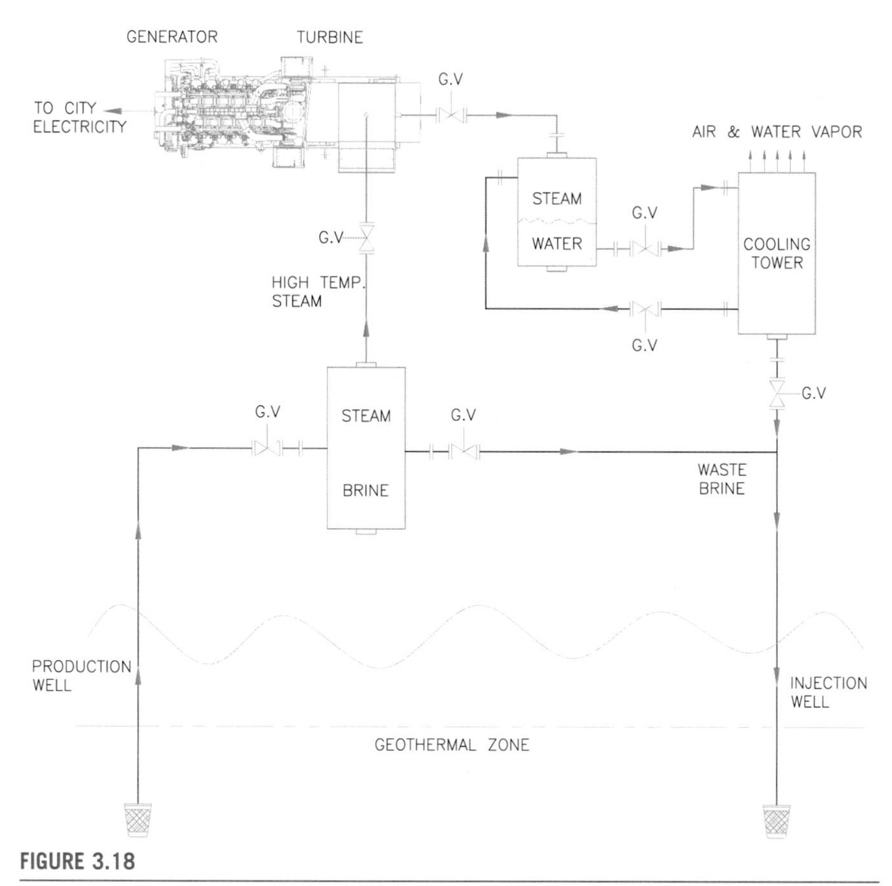

FIGURE 3.18

Binary steam geothermal power plant indicating the stages of primary and secondary fluids.

3.2.4 Solar power plants

Solar power plants utilize thermal energy from the sun, which is abundant, available, intermittent, yet cheap. This thermal energy is further transformed into electrical energy using photovoltaic panels. This is one type of solar power plants. Simply, a large number of panels are installed in an optimal configuration and harvest light energy from the sun and convert it into electrical energy which feeds into the grid. Another type of solar power plant is the concentrated solar power plant, which composed of mirrors or lenses that are stationed in an organized way to concentrate collected heat to one specific position. This heat is further utilized to power a steam turbine that generates electricity. However, the most common solar power plant is the traditional photovoltaic (PV) option. Solar capacity for each country varies depending on the solar irradiance as well as the available land. This type of power plant is considered a renewable option as the energy source is the sun, which is a clean, renewable, abundant, and cheap source. Solar PV farms can be ground mounted or roof mounted. Additionally, the ground-mounted systems can be fixed arrays, or installed with a single or a dual axis tracker. The modules are usually oriented toward the equator, with a tilt angle that is slightly lower than the site's latitude. Different tilt angles can be explored to find the optimal power production. Axis trackers are used to optimize performance as they allow for panels to track the sun as it moves position throughout the day. Fig. 3.19 shows an illustration that depicts a solar power plant. Once the thermal energy is harvested, solar panels convert it into direct current (DC) electricity. To convert this to alternating current (AC) electricity, another component becomes essential in the solar power plant, which is the inverter. There are different types of inverters including centralized and string inverters. Centralized inverters have higher capacity, in the order of 1 MW, while string inverters are significantly lower in capacity, normally in the order of 10 kW. Fig. 3.19 depicts a solar power plant with its main features and components.

Normally, solar power plants are constructed on wide-open spaces, constructing a solar farm, which produces a significant amount of electricity. This type of power plant fulfills the peaking demand, as it is a limited and intermittent source. Unless the storage option becomes sustainable and durable, this type of power plants will remain limited to peaking demand and not the base load demand. The performance of solar power plants is a function of climatic circumstances along with the quality of the equipment used in the system. Furthermore, locations with higher solar insolation yield higher electric production. Besides, solar systems' efficiencies also vary depending on the type of panels used. This conversion efficiency is critical as it impacts the overall efficiency of the system. Moreover, other system losses include losses between the DC output and the AC input.

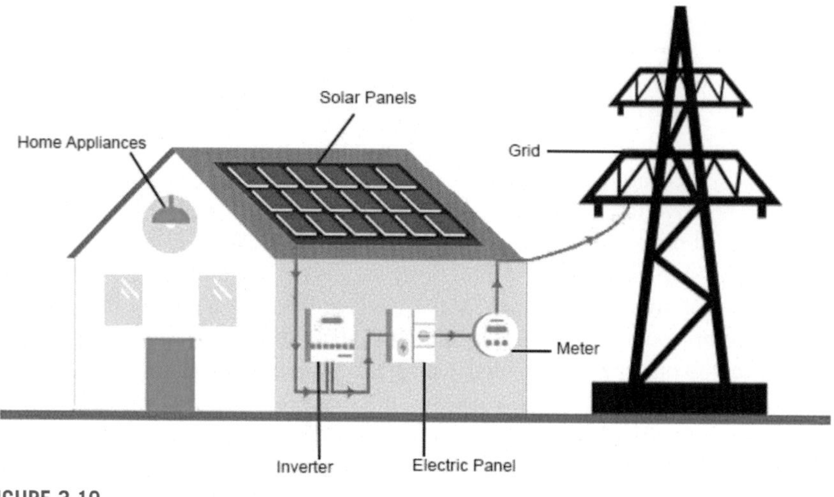

FIGURE 3.19

Rooftop solar system including rooftop panels, inverter, and grid integration.

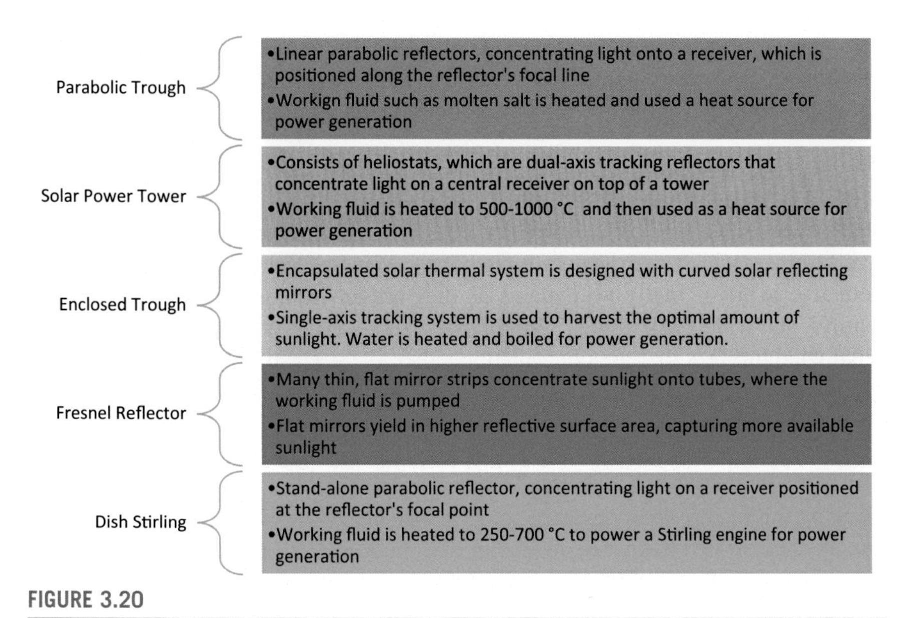

FIGURE 3.20

Various types for concentrators used in concentrated solar power plants.

Concentrated solar power (CSP) is another method to generate power using solar energy. After concentrating great amount of light into one source, heat is used to generate a steam turbine, which is connected to a generator to generate electricity. CSP is less common than PV plants, primarily because PV plants can still operate with cloud cover, while CSP is crucially impacted by any cloud cover. Furthermore, PV plant operating cost and production are much higher than that of CSP's. Moreover, the price per Watt from solar PV has significantly decreased, while system efficiency has increased, making power generation through this source somewhat lucrative. CSP uses various types of concentrators to yield different peak temperatures, which subsequently impact thermodynamic efficiencies. Fig. 3.20 shows different types of concentrators used for CSP.

Power generation through solar means has survived economically through various governmental incentives and grants such as feed in tariff, net metering programs, tax credits, and loan guarantees. The solar system's financial performance is a function of income and costs. Income is associated with the electrical output and the rate at which electricity is purchased. Although electricity prices may vary at times, support programs such as the ones aforementioned allow for sale rates to remain competitive and stable. As for the costs, the capital costs associated with solar power plants make up the dominant cost. Operating and maintenance costs are also considered when it comes to costs associated with this type of power generation.

Solar power plants can be off-grid and stand-alone systems or they can be connected to the grid in some capacity. Furthermore, these systems differ in size as some are simply for residential use and range between 6 and 10 kW while other solar farms may be massive in capacity ranging in MWs. Moreover, battery storage solutions are still underway with these types of power plants and could influence the market greatly if developed to the point of commercial competitiveness.

3.2.5 Wind power plants

Wind power plants rely on air flow from winds to provide mechanical power that drive turbines, which then generate electricity. Wind power is another renewable alternative to fossil fuel power plants as they are environmentally benign, clean, plentiful, and widely distributed. Wind power plants or farms consist of many individual wind turbines that are connected together to the electric grid. There are two types of wind power plants, onshore, and offshore. Fig. 3.21 shows the differences between these two types.

This type of power plant is very popular, and its capacity is increasing exponentially in the world. As these power plants rely solely on wind, kinetic energy is harvested by the blades and converted into mechanical energy that power an electric generator. Electricity produced feeds back to the local grid through a transformer for consumption. Fig. 3.22 illustrates an individual wind turbine and how electricity is generated.

The United Kingdom houses the largest wind farms in terms of capacity while the United States is the leading country when it comes to wind power production (IEA, 2018). Fig. 3.23 shows the world's top wind power producing countries along with their associated wind production in the year 2015.

3.2.6 Biomass power plants

Biomass power plants use animal and plant waste that cannot be used for feeding to provide energy. Waste such as horticulture and wheat stalks can be burnt to provide energy. Although burning this material causes carbon emissions, it is considered a renewable energy source if these stocks can be replaced with new growth. As this type of power plant is similar in process to coal power plants, many coal power plants get converted to biomass power plants for better efficiency and environmental

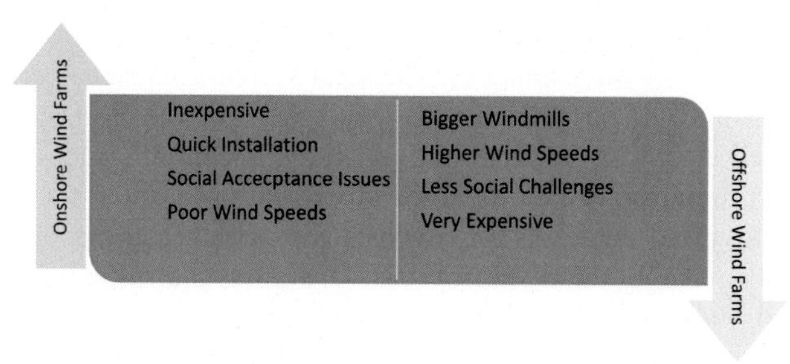

FIGURE 3.21

Types of wind farms including the main advantages and disadvantages.

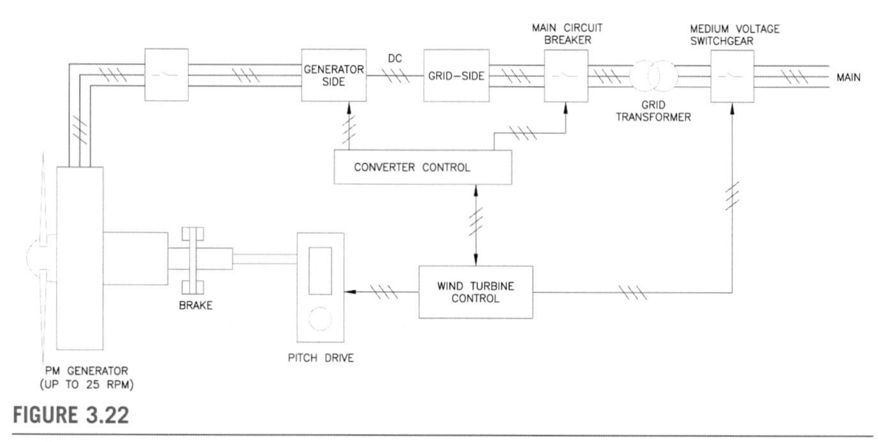

FIGURE 3.22

Schematic of an individual wind turbine and electricity generation.

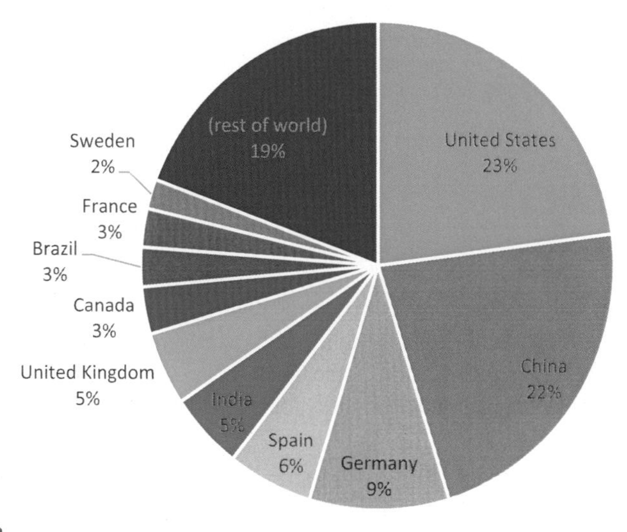

FIGURE 3.23

World's top producers of wind power in 2015.

Data from IEA, 2018.

considerations. Simply, the feedstock is fed into a massive burner, which burns the biomass. The resulting steam then powers an electric generator that produces electricity. Fig. 3.24 illustrates a simplified version of a biomass power plant. Biomass power generation can be achieved through various means. Combustion is the most common method for power generation when using biomass. As mentioned previously, this method allows for the burning of biomass directly in a boiler to provide steam that is further used to generate electricity.

On the other hand, gasification combustion is another method where solid biomass breaks down to form a flammable gas. Mixed burning is also used, where biomass is

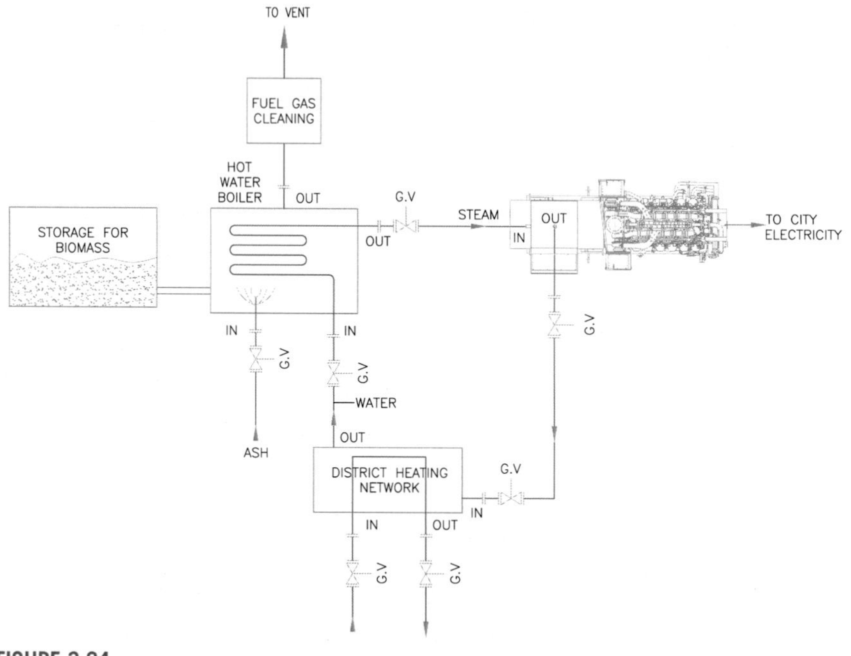

FIGURE 3.24

Biomass power plant and the process associated with electric generation.

mixed with coal and then burnt in the boiler together. Another method for processing biomass is the gasification mixed burning method. In this method, biomass is gasified first then the fuel gas is burnt with coal in the boiler. Fig. 3.25 lists the main advantages and disadvantages along with other features for each method.

3.3 Heating systems

Heating applications range widely from residential to industrial and from water to space heating. Heating systems are those systems that extract heat from various sources and convert it into a useful commodity such as space heating. Furthermore, these heating systems preserve temperatures within specific ranges by using thermal energy management within a structure. Heating systems could be classified as central or distributed systems. Central systems provide heat comprehensively to the overall space while heat is generated at one specific site.

Heat is transported through hot forced-air, which travels through ducts or hot water traveling through pipes, or by traveling steam through pipes as well. Various heating systems are summarized in Fig. 3.26. Similar to power generation, heating systems can be derived from various sources. The selection of the primary energy source for heating is dependent on cost, efficiency, convenience, and reliability. Fossil fuel burning for heating purposes is difficult to control automatically, however is

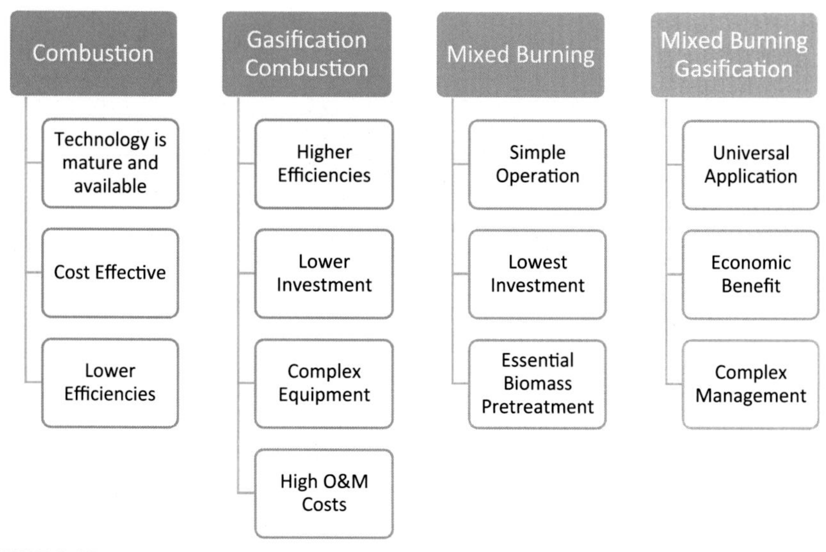

FIGURE 3.25

Biomass power plant generation methods and features of each method.

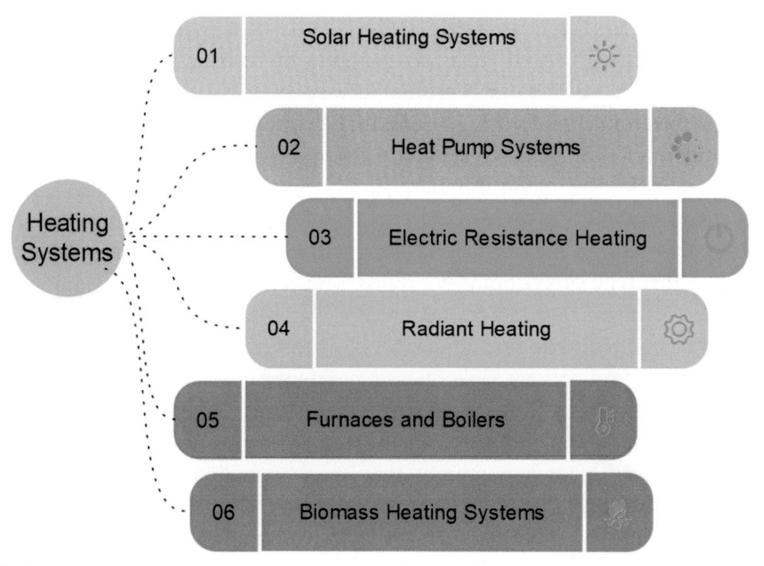

FIGURE 3.26

Various types of heating systems.

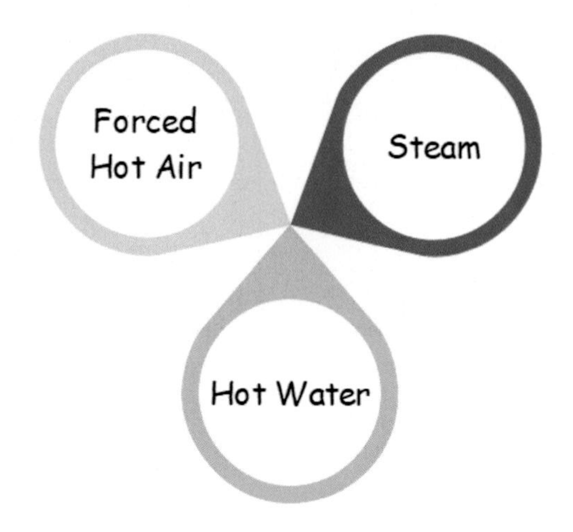

FIGURE 3.27

Main heat-transporting methods in central heating systems.

still a source that is frequently used. Coal on the other hand is phased out due to its detrimental environmental and health concerns. Liquid fossil fuels such as natural gas and kerosene are now widely used for heating applications. They can be automatically controlled and unlike wood pellets, do not require any ash or waste removal. In remote areas, where gas pipes are not serviced, propane tanks can be used as the primary source. In areas where electric cost is low and economically viable, electric heating is used via a heat pump or by resistance heating. Moreover, district heating systems feature centrally located boilers or water heaters, which transfer heat to various homes using a centralized circulation system. Central heating features three main methods as discussed earlier and illustrated by Fig. 3.27.

In the steam systems, central boilers supply steam at high pressures through a comprehensive distribution system, entrenched within the building envelope. Steam heating systems utilizes the high latent heat, which is released upon condensation. Furthermore, this type of system is equipped with a radiator in each room, coupled with a boiler. Moreover, such systems are not suitable for residential applications because of their high cost and maintenance. However, they are common in commercial and industrial applications. Water hearing systems, also known as hydronic heating systems run through a closed loop system. Water is heated through boilers, fueled by various types of fuels aforementioned. After heating, water is circulated through pipes to different locations as they pass by radiators to release the heat into the rooms. Because this is a closed loop system, the same water that is being heated is also reheated multiple times. Electric heating converts electric energy directly back to thermal energy. Normally, this type of system is expensive, relative to heat generated from natural gas or other types of fuels. However, if electricity rates are lower, it becomes a viable option for heating. Examples of this type of system include baseboard heaters and thermal storage systems. Heat pumps are also widely used for heating applications. This includes geothermal heat pumps and air source heat pumps. Geothermal heat pumps prove to be a more viable option as they extract stored heat from beneath Earth's surface and utilize it for space heating and domestic hot water heating. All of these types of heating systems have environmental aspects associated with them, as heat is lost throughout the process. Heat loss at the production, transportation, and recycling phases can be minimized when taking into consideration the system design, size, and architecture.

3.3.1 Solar heating systems

Solar heating systems can be divided into two groups, passive solar and active solar heating. In essence, these systems harvest thermal energy from the sun and utilize the collected heat for space heating purposes or to heat domestic water. Passive solar systems rely on the structure of the building to collect heat. This could be in the form of a tilt or a roof orientation that allows for higher solar irradiance. On the contrary, active solar heating systems rely on heat pumps that transfer the collected heat from the solar collectors to the building. In contrast to photovoltaic panels that generate electricity, thermal solar panels are used to capture energy from the sun and utilize it to provide the abovementioned commodities. Fig. 3.28 illustrates an example of a solar heating system. Normally these systems are not designed for stand-alone systems due to the intermittency of the energy source and the lack of viable storage options. However, these systems can be complimentary to other existing heating systems powered by natural gas or other sources.

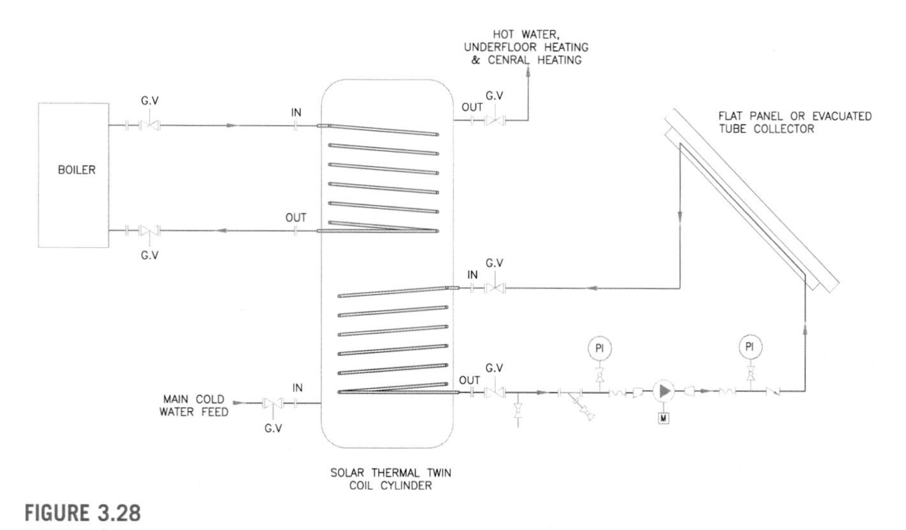

FIGURE 3.28

Schematic of a solar water heating system and all of the subcomponents.

3.3.2 Geothermal heating systems

Geothermal heating systems utilize underground temperatures of the Earth to heat domestic water or for space heating. Earth gets heated by the sun, as this heat is stored underground and in the water. The earth's absorption of the heat from the sun yields in consistent ground temperatures ranging from 6 to 38 °C depending on geographic location. Therefore, geothermal heating systems are composed of a series of underground pipes, which could be vertical or horizontal, with water, steam, or a refrigerant circulating and transporting heat from underground to the building envelope. Fig. 3.29 illustrates an example of a geothermal heating system. Although these systems provide heat in the winter, they also have the capability to provide coolness in the summer by the reversing the process, where heat emitting from the building envelope is returned back underground and stored for winter months.

3.3.3 Biomass heating systems

Biomass heating systems utilize heat generated from various biomass processes to heat domestic water or provide space heating. Processes including direct combustion, gasification, CHP, and anaerobic and aerobic digestion can all be used to provide heat. These types of heating systems are useful as they use residues and wastes from agricultural, forest, industrial, and urban sectors to produce useful and necessary commodities. However, combustion for example has a negative environmental impact as it introduces significant air pollutants. Fig. 3.30 shows the various types of biomass heating systems depending on process and automation.

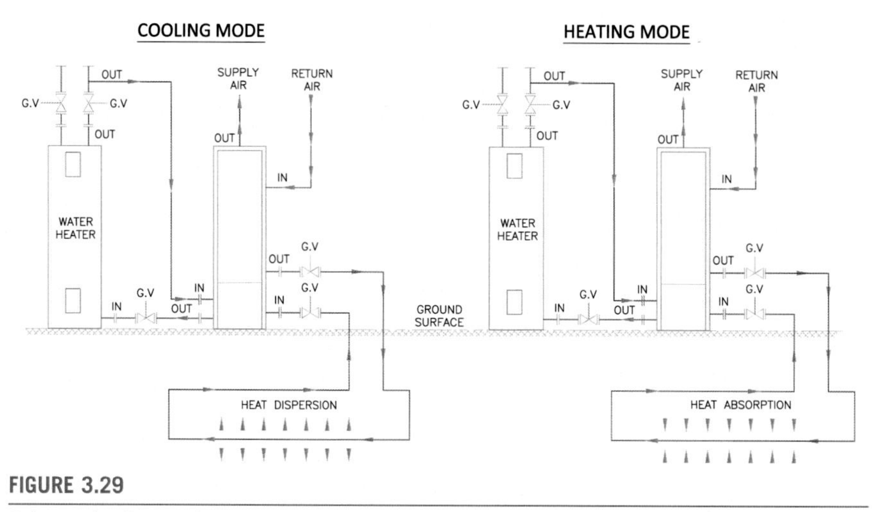

FIGURE 3.29

Schematic of a geothermal water heating and cooling system and all subcomponents.

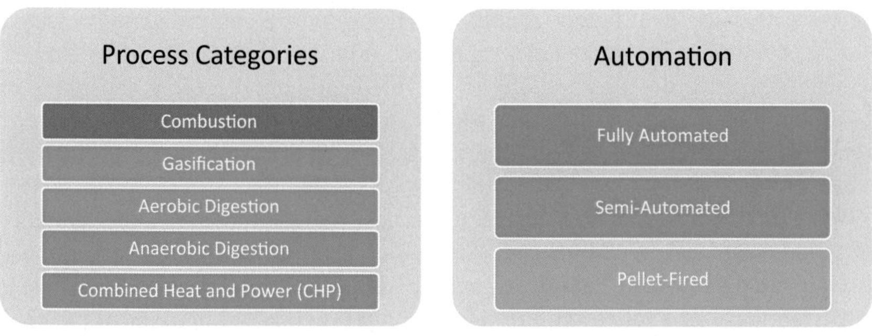

FIGURE 3.30

Types of biomass heating systems depending on different parameters.

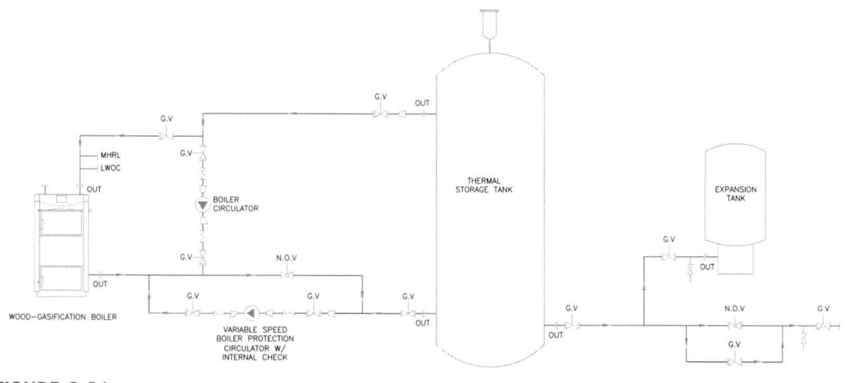

FIGURE 3.31

Biomass heating plant and the process associated with heat generation.

Hikes in fuel prices including oil, natural gas, and fossil fuels overall produced biomass heating systems, as they became more economic and competitive solutions. Buffer tanks store the hot water, acting as a thermal storage medium. This component is essential for the efficiency of these systems. In fact, details about the biomass heating system process and components are illustrated in Fig. 3.31.

3.3.4 Heat pumps

Heat pumps are devices that are powered by electricity or external power source, which transfer thermal energy from one source to another. Heat pumps consist of four main components including a condenser, expansion valve, compressor, and evaporator. A refrigerant, which is the heat medium that runs between these components, is the heat carrier. The performance of heat pumps is measured by the coefficient of performance (COP), which is equivalent to the system's percentage efficiency. Because heat pumps absorb heat from outside and circulate it inside a building envelope, it is considered a vapor-compression refrigeration cycle. Reversible heat pumps feature a dual ability to provide heating and cooling depending on the season. This duality in function results from employing a reversing valve to reverse the refrigerant's flow from the compressor to the condenser and finally to the evaporator.

Water is the most common refrigerant in heat pumps; however, other common refrigerants also include R134a and Ammonia. Thermal energy transportation is achieved through flowing liquid or gas. Air usage is impractical and inefficient. The performance of a refrigerator or a heat pump is expressed in terms of the coefficient of performance, which can be illustrated as follows:

$$COP_R = \frac{Q_L}{\dot{W}_{in}} = \frac{1}{\frac{Q_H}{Q_L} - 1} \tag{3.1}$$

$$COP_{HP} = \frac{Q_H}{\dot{W}_{in}} = \frac{1}{1 - \frac{Q_L}{Q_H}} \tag{3.2}$$

According to the Carnot principles, the thermal efficiencies of all reversible heat engines operating between the same two reservoirs are the same. Therefore, the following equations can be derived:

$$\left(\frac{Q_H}{Q_{L\,rev}}\right) = \frac{T_H}{T_L} \tag{3.3}$$

Therefore, the ratio of $\frac{Q_H}{Q_L}$ can be replaced by $\frac{T_H}{T_L}$ for reversible devices. Carnot heat engines are ones that can operate on the reversible Carnot cycle. The thermal efficiency of a Carnot heat engine or any reversible heat engine is given by

$$\eta_{th,rev} = 1 - \frac{T_L}{T_H} \tag{3.4}$$

Furthermore, the COPs of reversible refrigerators and heat pumps are given in a similar fashion as follows:

$$COP_{R,rev} = \frac{1}{\frac{T_H}{T_L} - 1} \tag{3.5}$$

$$COP_{HP,rev} = \frac{1}{1 - \frac{T_L}{T_H}} \tag{3.6}$$

Furthermore, the maximum power that can be produced is given by

$$\dot{W}_{\text{rev}} = \eta_{\text{th,rev}} \times \dot{Q}_{\text{in}} = \left(1 - \frac{T_{\text{L}}}{T_{\text{H}}}\right) \times \dot{Q}_{\text{in}} \tag{3.7}$$

Moreover, exergy efficiency can be derived using the following equation:

$$\psi = \frac{\eta_{\text{th}}}{\eta_{\text{th,rev}}} \tag{3.8}$$

It can also be expressed as the ratio of the useful work output and the maximum possible reversible work output as such:

$$\psi = \frac{W_{\mu}}{W_{\text{rev}}} \tag{3.9}$$

Therefore, the exergetic COPs for heat pumps and refrigeration cycles can be expressed as follows:

$$\psi = \frac{\text{COP}}{\text{COP}_{\text{rev}}} \tag{3.10}$$

Fig. 3.32 shows the main components of a heat pump and the different state points.

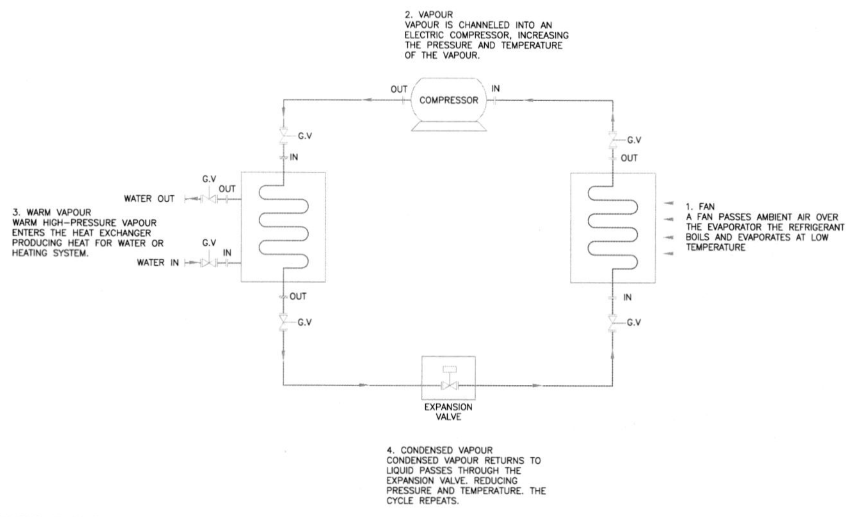

FIGURE 3.32

Heat pump cycle with the four main components.

3.4 Refrigeration systems

Refrigeration or cooling systems simply shift the heat from reservoirs with lower temperatures to reservoirs with higher temperatures. In fact, it is a reverse process of the heat pump, thus providing cooling applications. Four types of refrigeration cycles that are most commonly used are summarized in Table 3.2.

In the vapor-compressor cycle, refrigerant enters the compressor as a saturated vapor at state 1. The refrigerant undergoes isentropic compression to match the pressure of the condenser. It then exits the compressor toward the condenser as a superheated vapor at state 2. The refrigerant is cooled at the condenser and transformed back to liquid at state 3. During this process, heat is lost from the condenser. The refrigerant then passes through an expansion valve, where the pressure is dropped significantly. Finally, the refrigerant reaches the evaporator at state 4, where it draws heat, causing refrigerant to vaporize again and continue the cycle back at the compressor. Therefore, the four processes of the ideal vapor-compression refrigeration cycle are as follows (Çengel et al., 2019):

1-2 Isentropic compression in a compressor
2-3 Constant-pressure heat rejection in a condenser
3-4 Throttling in an expansion device
4-1 Constant-pressure heat absorption in an evaporator

Furthermore, Fig. 3.33 illustrates the refrigeration cycle in detail.

The vapor-absorption cycle is very similar to the vapor-compression cycle, apart from the pressure increase of the refrigerant. The cycle also features the absorption of the refrigerant by a transport medium. Moreover, the most common vapor-absorption refrigeration system is the ammonia-water system. In this system, the compressor is replaced by an absorber as shown in Fig. 3.34. The absorber causes

Table 3.2 Different types of refrigeration systems and their description.

Refrigeration system	Description
Mechanical-compression refrigeration systems	Refrigerant compresses mechanically to low pressures and then expanding it to high pressures. Temperature difference is utilized for cooling applications.
Absorption refrigeration	Refrigerant and absorbent are mixed and further processed to extract heat and they carry on the refrigeration cycle to provide cooling.
Evaporative cooling	Does not use traditional refrigeration cycle. Rather, water absorbs the heat from the air and evaporates. Cooler air is rerouted to the building envelope.
Thermoelectric refrigeration	Neither refrigerant nor water is used in this process. Uses electric current and a thermocouple. Current direction is reversed with the cold side directed at the area to be cooled.

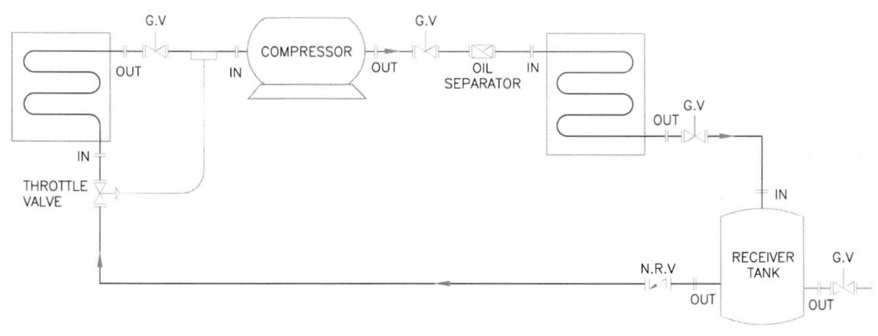

FIGURE 3.33

Vapor-compression refrigeration cycle with the four main components.

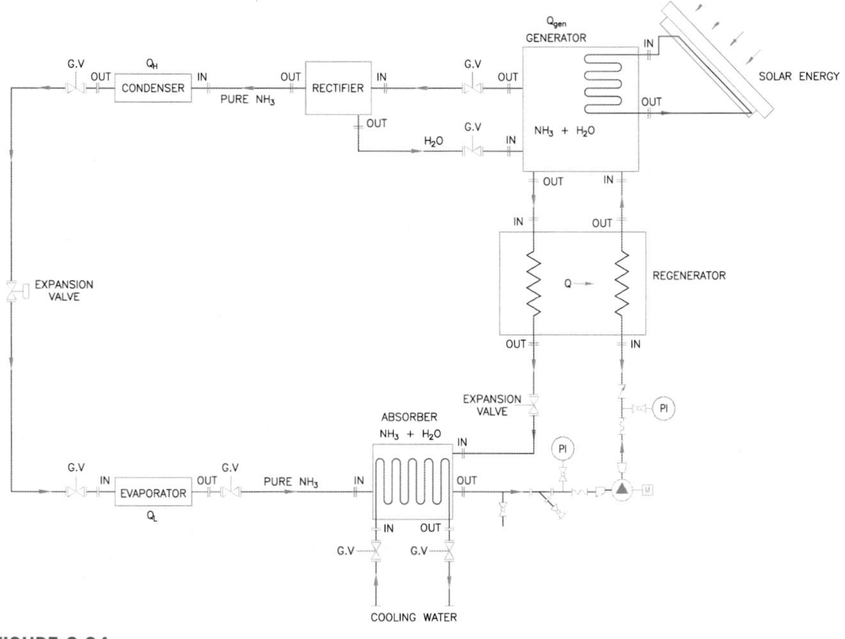

FIGURE 3.34

Vapor-absorption refrigeration cycle with the main components.

the refrigerant to dissolve in a suitable liquid. The absorber also allows for the ammonia's pressure to increase and then it is cooled at the condenser with heat losses being emitted. Thereafter, as it throttles through the evaporator, it absorbs heat from the refrigerated space. The work input for a vapor-absorption refrigeration system is much lower than the work input for a vapor-compression refrigeration system. This is because a liquid is being compressed instead of a vapor.

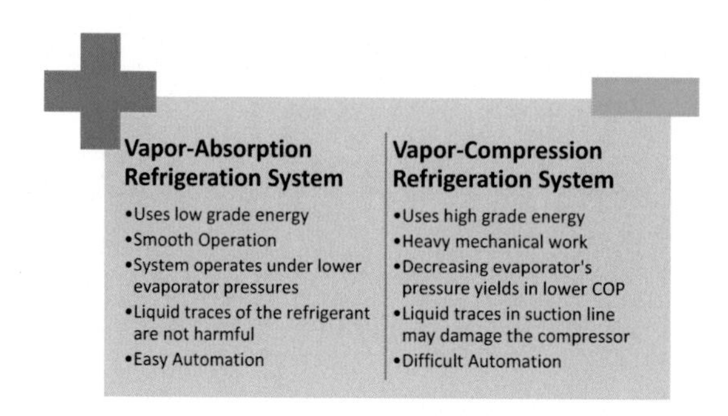

FIGURE 3.35

Comparison between vapor-absorption and vapor-compression refrigeration systems.

However, these systems tend to be expensive, more complex, occupy larger space, and are less efficient than the vapor-compression refrigeration systems. Fig. 3.35 compares the vapor-compression and vapor-absorption refrigeration systems. The main difference between vapor-compression and vapor-absorption refrigeration systems is that vapor-compression uses higher-grade energy and requires heavy mechanical work whereas the vapor-absorption system uses lower-grade energy and results in smoother operation.

Therefore, they are only preferred when a heat source is more readily available that electricity. In fact, the integration of such systems to harvest waste heat from industrial processes or to utilize solar thermal energy is most beneficial.

Although the vapor-compression refrigeration cycle denotes a reverse Rankine cycle, the gas refrigeration cycle is a reverse Brayton cycle. In these types of refrigeration systems, gas is the working fluid and is compressed or expanded, but does not undergo any phase changes. In this system, gas is compressed at the compressor, reaching very high temperature and pressure. The gas is then cooled at constant pressure until reaching ambient temperature at the high-grade heat exchanger. The gas is then expanded through the turbine, and the temperature drops. Because the gas refrigeration cycle is not isothermal, it is distinct from the reversed Carnot cycle. Therefore, these systems have lower COPs than the vapor-compression refrigeration systems. Nonetheless, these systems are desirable because of their simplicity, lighter components, making them ideal for aircraft cooling for instance. Fig. 3.36 illustrates the gas refrigeration cycle.

3.5 Refineries

Refineries are complex production facilities comprising various energy processes that utilize raw materials to useful commodities or products. Petroleum oil refineries and natural gas processing are the two most common types of refineries associated with energy production and consumption. In a petroleum oil refinery, crude oil is processed through various subprocesses until reaching the final useful commodities

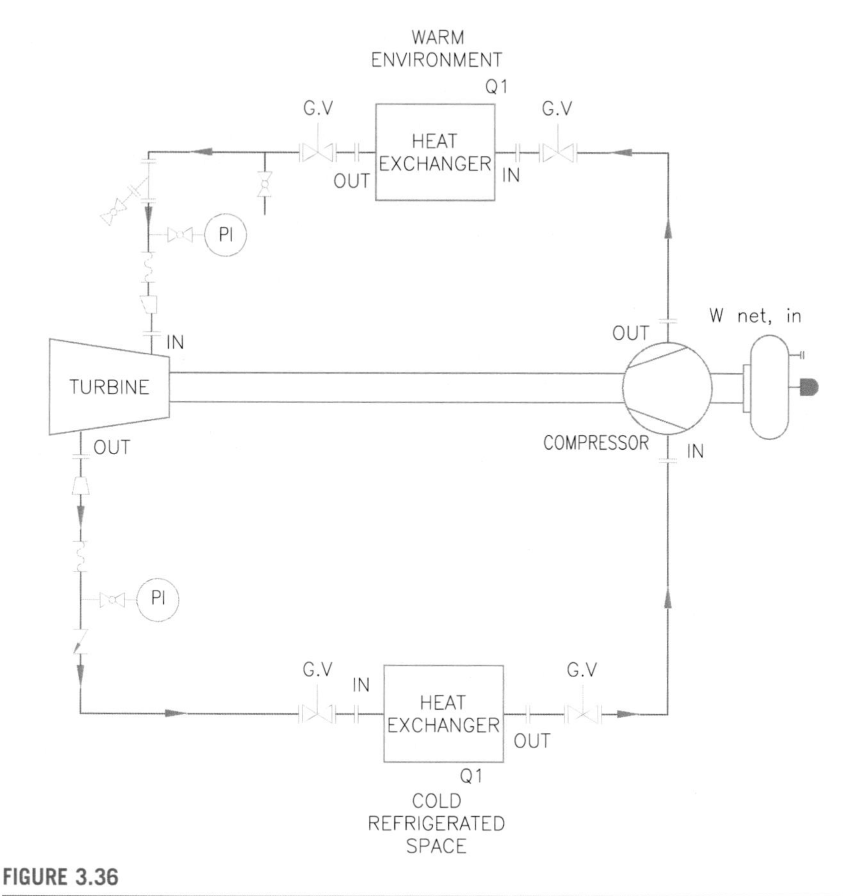

FIGURE 3.36

Gas refrigeration cycle with the main components.

such as gasoline or asphalt. Furthermore, a distillation column is the primary component for any refinery. In this column, crude oil at various temperatures, pressures, and boiling points is divided into various petroleum products. The most basic type of refinery is also called topping refinery, which consists of a simple distillation column only. It indeed produces naphtha. Hydroskimming refineries produce gasoline in addition to naphtha and are more complex than topping refineries. In addition, cracking refineries are equipped with vacuum distillation and catalytic cracking, which enables the production of light and middle distillates. Furthermore, coking enables the processing of vacuum residue into useful products. Finally, an integrated refinery has the ability to upgrade naphtha and liquid petroleum gas into their basic petrochemicals by producing benzene or naphtha cracking for instance. Fig. 3.37 illustrates the different types of refineries and their associated products.

Although crude oil can be further processed to retrieve a wide range of useful fuels and products, the major fuel extracted is gasoline as illustrated by Fig. 3.38.

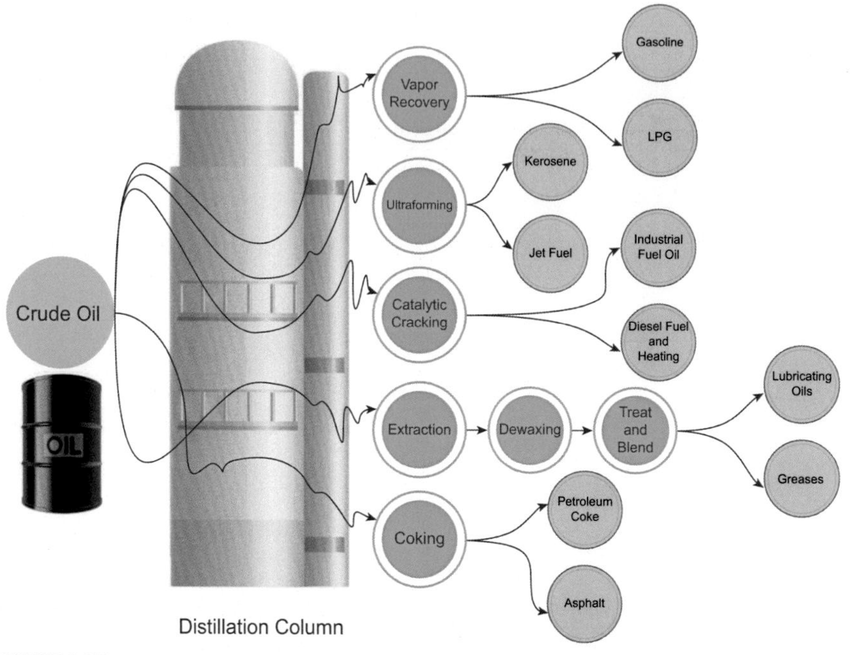

FIGURE 3.37

Typical oil refinery with the main processes and products.

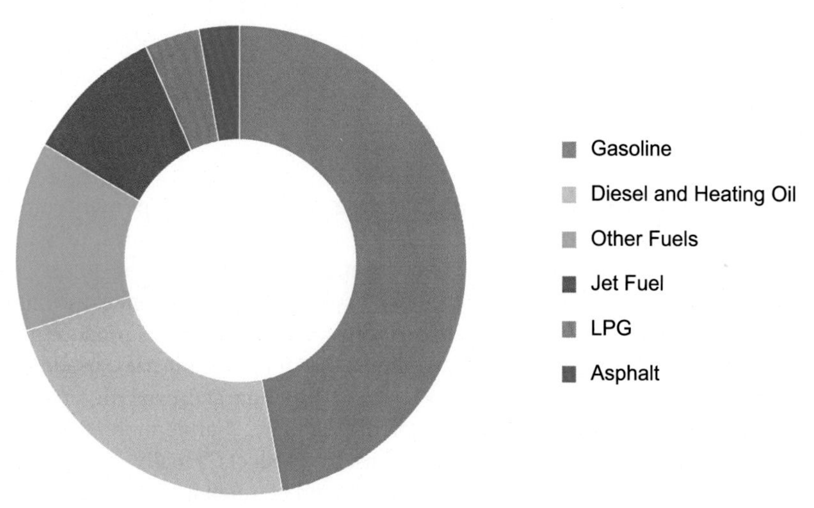

FIGURE 3.38

Breakdown of the products retrieved from a typical oil barrel.

Data from US Department of Energy, 2011.

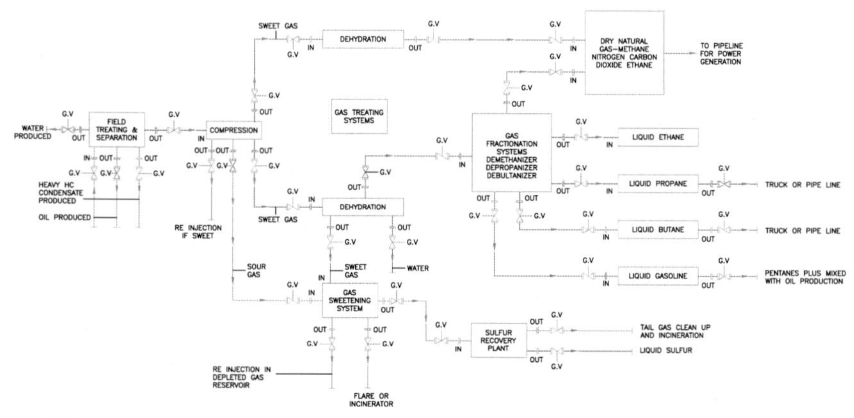

FIGURE 3.39

Typical natural gas processing plant with the main processes and products.

In fact, gasoline represents approximately half of a typical American oil barrel. Diesel, jet fuels along with other fuels make up another 40% of the barrel, while LPG and asphalt make up around 7% of a typical oil barrel. Petroleum refineries have economic, political, and social implications and thus, their existence is dependent on many factors. Furthermore, oil refineries are a combination of very complex processes that require substantial amounts of heat and electricity. This energy is either produced onsite or purchased externally.

On the other hand, natural gas processing is also a complex industrial process, which is intended to purify natural gas from any impurities, yielding in high-quality natural gas. Depending on the depth, the geologic formation, and the location of underground deposits, the quality of natural gas varies. Normally, reservoirs containing oil also contain natural gas. Common contaminants to natural gas include water, carbon dioxide, and hydrogen sulfide. Associated gas is a term given to natural gas found along with oil reservoirs, usually acting as a cap or sometimes dissolved within the oil reservoirs. On the other hand, nonassociated gas is given to the natural gas found where there is no crude oil. Fig. 3.39 illustrates the flow of natural gas processing.

3.6 Closing remarks

In conclusion, energy systems utilize various primary energy sources and energy conversion resources to provide necessary and useful commodities. This chapter provided a thorough explanation about the various types of power plants including fossil fuel power plants and renewable-based power plants. Other cooling and heating systems including district heating and absorption chiller systems are also discussed in detail in this chapter. Furthermore, nuclear 63% of all nuclear reactors worldwide are pressurized water reactors, which use water as a moderator and enriched uranium oxide as fuel, with an overall system cycle efficiency of 32%. Moreover, geothermal plants can be optimized by adding more flash separators to enhance the efficiency. Roughly, each extra flash separator can increase the overall efficiency by 10%–15%. In addition, wind energy is dominated by the United States and China, accounting for approximately 45% of the global wind energy production.

CHAPTER

Energy services

4

4.1 Introduction

Energy services are considered the end step after energy sources and energy systems. The pathway from energy sources to energy services go through storage mediums as well. For example, storage is vital between the energy source and energy system stage. Similarly, storage is equally important between the energy system and energy service stage. As mentioned earlier, there are various types of commodities that can be desired. A list of commodities is summarized in Fig. 4.1. The main services such as electricity and space heating are considered essential for residential purposes while other services such as drying is used widely in the industrial sector. These services such as electricity come from various sources such as nuclear and fossil fuels.

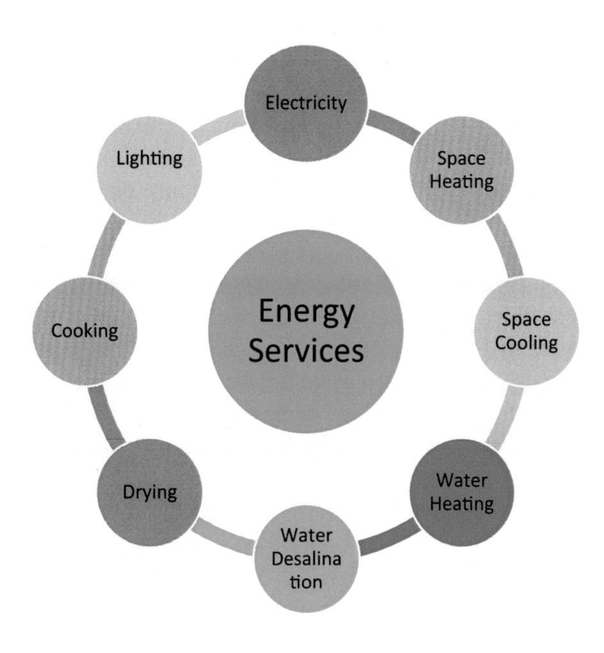

FIGURE 4.1

Examples of various types of energy services and commodities.

Energy Sustainability. https://doi.org/10.1016/B978-0-12-819556-7.00004-8

Similarly, heating and water desalination is also achieved using various sources such as coal and oil. Renewable energy sources can be utilized for lighting applications. Although electricity cannot be stored, potential energy can be stored in various intermediate forms using thermal energy storage and other storage solutions. Furthermore, numerous energy systems can be designed to produce these commodities. For instance, electricity can be produced from fossil-based fuels such as coal and oil using steam turbines and organic Rankine cycles or from nuclear through atomic fusion, or from renewable sources such as solar, wind, and hydro.

Two of the most important commodities among all of the ones listed earlier are clean water and electricity. Electricity can be generated through various means and is distributed in an organized way through an interconnected grid system. Rural areas away from metropolitan concentrations are sometimes too far for the grid to reach to them and thus they may be off grid.

4.2 Electricity

Electric power refers to the rate of energy transfer from an electric circuit. This electric energy is measured in the unit of watt or joule per second. Electricity is used in various applications as we evolve into a digital world that is primarily powered through electricity and smart energy solutions. Electricity is used for lighting, transportation, powering essential home appliances, machineries, and systems. In essence, electric charges are of two types: positive and negative charges. These electric charges either pull or push each other if they are not touching due to the electric field created by each charge around itself. Similar to magnets, electricity creates a magnetic field, where similar charges repel each other while opposite charges attract each other. Therefore, two negative charges result in repulsion while opposite charges attract. Furthermore, while some electrons can be stuck to certain objects, other electrons can move around the material depending on its type. Protons, on the other hand, never move around objects due to its heavy weight. Conductors are materials that allow electrons to circulate around it. On the other hand, insulators are materials that keep electrons limited to a specific place. Conductors include aluminum, copper, silver, and gold. Insulators include plastic, rubber, and wood. The conductive capacity for each metal varies depending on the physical properties.

Global electricity demand is forecasted to rise considerably to 38,700 TWh by 2050. This is significant considering the demand of 25,000 TWh in 2017. Fig. 4.2 shows the change in gross electricity demand for various countries and continents for various periods.

Processes such as self-consumption of power plants, grid losses, and storage losses account for 17% of the total electricity production. The world's total electricity generation accounted for 20,261 TWh in 2008 and only 16,816 TWh made it to final consumption. Table 4.1 shows the electricity consumption by category for 2008 for various countries.

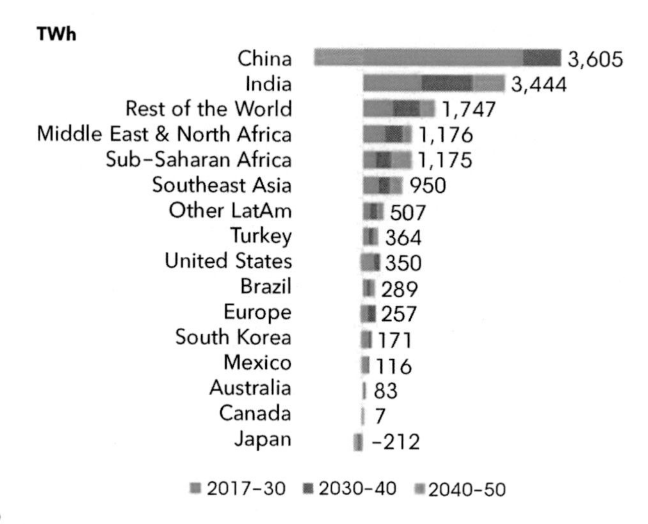

FIGURE 4.2

Global change in electricity demand for various durations (Bloomberg NEF, 2018).

4.3 Heating and cooling

Heating and cooling commodities are critical energy services that are needed seasonally. Space heating makes up the major energy consumption in the winter for countries like Canada. These services encompass a wide range of applications and technologies such as cooking, water heating, ambient cooling, and refrigeration. Heating systems could come in various forms including block heaters, where the engine is heated using electric heating to ease the starting of the engine during winter months. On the other hand, cathode heaters utilize coils or filaments that are used to heat the cathode in a vacuum tube or cathode ray tube. Central heating is another method for heating, where heat is distributed from one point to various rooms or units within a building. A convector heater is another type of heating system that operates by utilizing air convection currents circulation through a body of an appliance. As discussed before, district heating systems are also efficient systems where heat is generated from one centralized location for residential and commercial heating applications. Other types of heating systems include feedwater heating, fireplace, gas heating, geothermal heat pumps, heating pads, hydronics, induction heating, radiant heating, radiators, solar furnace, solar heating, and water heating.

The demand for heating in the residential and industrial sectors exceeds the demand for cooling. However, the cooling demand is also increasing with climate change development. Fig. 4.3 shows the renewable heat consumption by country between 2017 and 2023.

Heat can be generated through various means. Renewable heat consumption is expected to grow up to 20% between 2018 and 2023 (IEA, 2018). Half of the generated heat is consumed by the industrial sector for processes such as drying, process

Table 4.1 Total electricity consumption for various countries based on sector.

Country/ geographical region	Total consumption (TWh)	Industry	Transport	Commercial	Agriculture	Fishery	Residential	Other
China	2842	67.80%	1.05%	5.40%	3.12%	0.00%	15.50%	7.19%
India	602	46.40%	1.93%	8.00%	17.92%	0.00%	20.70%	5.05%
USA	3814	24.00%	0.20%	35.00%	0.00%	0.00%	36.20%	4.59%
Indonesia	129	37.20%	0.00%	23.90%	0.00%	0.00%	38.90%	0.00%
Brazil	410	48.10%	0.39%	23.70%	4.49%	0.00%	23.30%	0.00%
Pakistan	70	27.50%	0.01%	14.20%	12.50%	0.00%	45.90%	0.00%
Bangladesh	32	56.30%	0.00%	6.00%	3.37%	0.00%	32.90%	0.00%
Nigeria	19	20.00%	0.00%	24.70%	0.00%	0.00%	55.30%	0.00%
Russia	725	49.60%	11.45%	20.60%	2.14%	0.04%	16.10%	0.00%
Japan	964	31.50%	1.95%	36.40%	0.09%	0.00%	29.80%	0.23%
Mexico	200	61.30%	0.55%	10.30%	4.05%	0.00%	23.70%	0.00%
Philippines	49	34.60%	0.23%	28.70%	2.30%	0.31%	33.80%	0.00%
Vietnam	68	51.80%	0.75%	8.10%	0.97%	0.00%	38.40%	0.00%
Ethiopia	3.1	38.00%	0.00%	23.60%	0.00%	0.00%	37.70%	0.74%
Egypt	112	33.40%	0.00%	15.40%	4.13%	0.00%	39.20%	7.84%
Germany	526	46.10%	3.14%	22.60%	1.66%	0.00%	26.50%	0.00%
Turkey	159	45.40%	0.60%	25.60%	3.54%	0.10%	24.80%	0.00%

DR Congo	6.1	63.40%	0.00%	3.10%	0.00%	0.00%	33.50%	0.00%
Iran	164	33.20%	0.15%	19.00%	12.92%	0.00%	32.30%	2.50%
Thailand	135	42.40%	0.04%	35.60%	0.21%	0.00%	21.30%	0.54%
France	433	32.60%	3.06%	25.00%	0.88%	0.03%	35.90%	2.57%
UK	342	33.20%	2.47%	28.60%	1.19%	0.00%	34.50%	0.00%
Italy	309	45.80%	3.50%	26.80%	1.81%	0.02%	22.10%	0.00%
South Korea	407	51.00%	0.55%	32.50%	1.61%	0.45%	13.80%	0.00%
Spain	265	38.90%	1.10%	29.50%	2.29%	0.00%	27.10%	1.08%
Canada	519	36.30%	0.81%	30.00%	1.86%	0.00%	31.00%	0.00%
Saudi Arabia	170	12.40%	0.00%	28.50%	2.04%	0.00%	56.90%	0.14%
Taiwan	210	55.70%	0.52%	13.70%	0.78%	0.46%	20.30%	8.48%
Australia	212	44.70%	1.33%	25.60%	0.88%	0.00%	27.40%	0.00%
Netherlands	109	38.60%	1.48%	30.00%	7.15%	0.00%	22.70%	0.00%
World	**16,816**	**41.70%**	**1.60%**	**23.40%**	**2.50%**	**0.03%**	**27.40%**	**3.43%**

Data from IEA (2018).

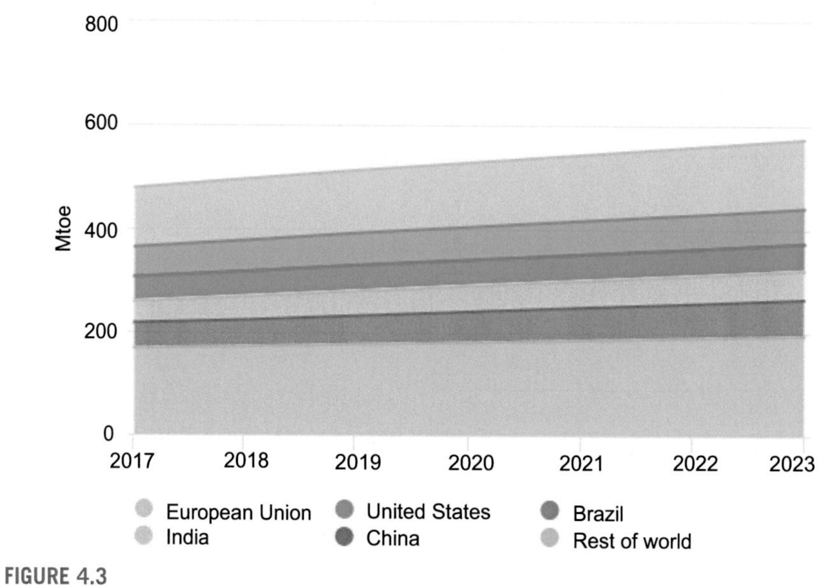

FIGURE 4.3

Global renewable heat consumption and forecast between 2017 and 2023 (IEA, 2018).

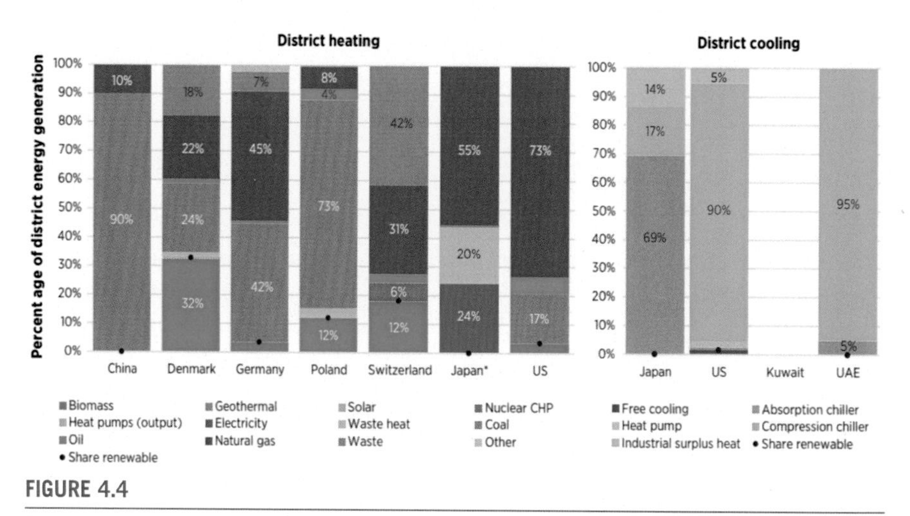

FIGURE 4.4

District heating and cooling generation mix for selected countries (IEA, 2018).

heating, and hot water utilization. The residential sector also uses heating for space heating and hot water uses. Moreover, district heating and cooling generation and distribution various from country to country and from source to source. Fig. 4.4 depicts a comprehensive illustration of current district heating and cooling generation mix in selected countries.

Cooling is the energy transfer of thermal energy using thermal radiation, heat transfer, convection, and conduction. Devices used for cooling include heat exchangers, radiators in automobiles, intercoolers, coolants, cooling towers, heat sink, or heat pipe. Cooling towers rejects waste heat by cooling the incoming water into lower temperatures. Evaporation is used to remove process heat and cool the working fluid. Waste heat can also be utilized into further other processes using a multigeneration system and achieving cooling as well.

4.4 Closing remarks

In conclusion, energy services are the energy products and useful commodities delivered after processing energy sources via energy systems. The 3S rule is imperative in this structure as it characterizes energy based on sources, systems, and services with storage solutions in between each step. The development of viable energy storage solutions is critical in adopting environmentally benign energy sources and systems. Energy services such as heating, cooling, and electricity are now considered essential commodities for all sectors including residential, commercial, institutional, and industrial sectors. On the other hand, drying is a commodity that is specific to the industrial sector. All in all, energy consumption varies from country to country based on population, lifestyle choices, and economic and industrial productivity.

CHAPTER

Community energy systems

5

5.1 Introduction

Community energy systems are becoming more popular and efficient than conventional energy systems. This is because they are decentralized, modular, and more flexible than the traditional energy systems. Furthermore, these systems are usually located near the communities they serve to avoid long transmission of energy. On the other hand, these systems are limited in their capacity. Moreover, community energy systems or distributed energy systems are usually hybrid systems, comprising multiple generation and storage units. Similarly, district heating systems comprise a network of pipes that deliver heat to residential and nonresidential consumers. The distributed heat is generated in centralized locations through cogeneration plants for instance. District heating plants feature higher efficiencies and better environmental impact than localized boilers. Furthermore, microgrids are localized, modern, and small-scale grids that are able to function with full independence from the local grids as well as strengthen the grid's resilience and help to mitigate grid disturbances. Moreover, microgrids are typically low-voltage AC grids that use various sources for power generation including solar systems and hybrid systems. These small community energy systems have an advantage of being near the cities they serve. This is indeed a benefit, as major energy systems cannot be close to cities such as coal plants due to the severe environmental footprint or hydroelectric systems due to their site limitation. Fig. 5.1 illustrates the various types of community energy systems.

5.2 Combined heat and power

This energy system refers to the production of two useful commodities from the same process. In this case, electricity and heat are generated simultaneously using a cogeneration power plant. This type of energy system is more efficient as it utilizes the waste heat from the power generation process and produces useful heat. Furthermore, by-product heat from this system can also be further used in absorption refrigerators through a trigeneration system. Moreover, in these systems, high-pressure steam is used for power generation first, followed by lower pressure steam, which

Energy Sustainability. https://doi.org/10.1016/B978-0-12-819556-7.00005-X

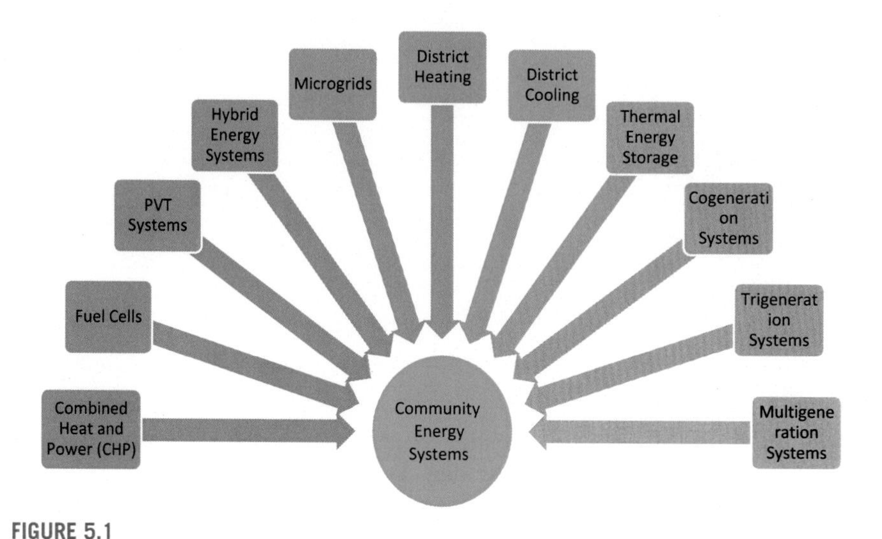

FIGURE 5.1

Various types of community energy systems.

is used for space or water heating. Although this system requires substantial capital cost at first, it is one of the most cost-efficient methods in the long term and more environmentally friendly. Fig. 5.2 shows the various types of combined heat and power (CHP) plants.

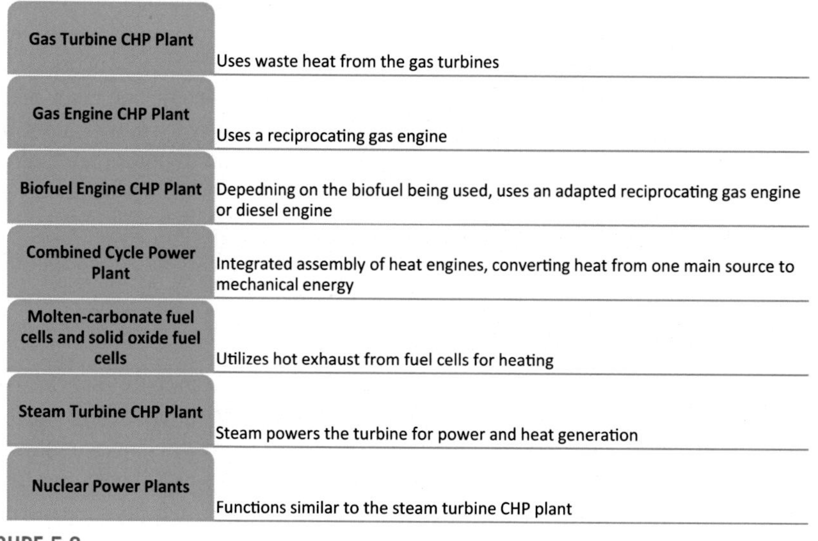

FIGURE 5.2

Types of combined heat and power energy systems.

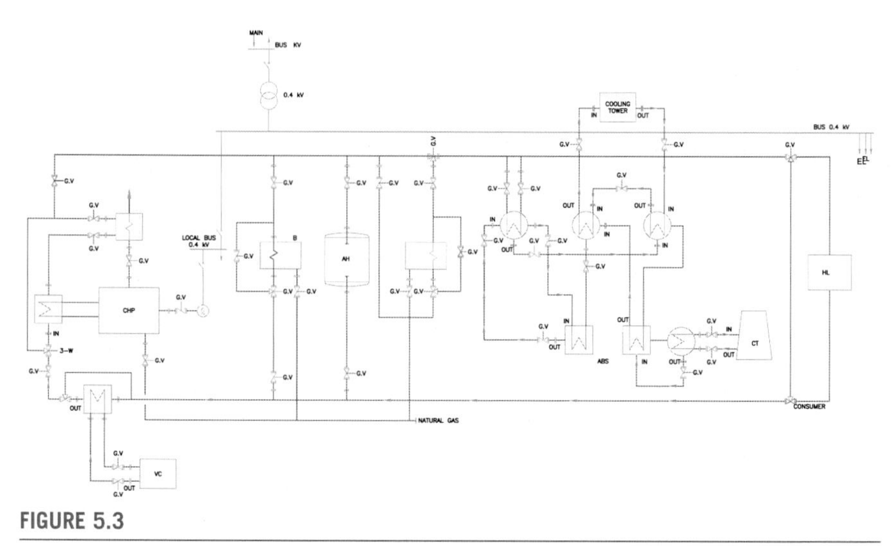

FIGURE 5.3

Gas engine CHP plant including main components and processes.

Although some CHP plants may use Stirling engine or reciprocating engine, others could use biomass or solid waste as the burning fuel. Furthermore, CHP systems usually yield higher thermal efficiencies due to the additional useful output they provide with relation to the total heat input. In industrial applications, excess heat is recycled for other duties such as drying, or steam processing. Moreover, this type of energy system has proved practical benefit to industries, resulting in reduced cost and environmental footprint. In fact, power can be generated on-site instead of importing it from the grid. Furthermore, the ability for power generation on-site enables industries to be independent of local utilities, including price fluctuations depending on the time of use as well as regular challenges that utilities face such as shutdowns, and regular maintenance. Fig. 5.3 illustrates a sample CHP power plant including all the main components and processes.

5.3 Fuel cells

Fuel cells are electrochemical cells that have the ability to produce electricity given a continuous supply of fuel and oxygen. In the cell, an electrochemical reaction occurs between hydrogen fuel and oxygen. Although fuel cells vary in type, the basic structure of a fuel cell consists of a cathode, an anode, and an electrolyte. Oxidation reaction is triggered by a catalyst at the anode, thus generating protons and electrons. This reaction also allows for the flow of protons through the electrolyte from the anode to the cathode.

Simultaneously, direct current electricity is produced by drawing electrons from the anode to the cathode through an external circuit. In addition, water is formed at the cathode due to the reaction between hydrogen ions, electrons, and oxygen. The type of electrolyte used in the fuel cell determines the type of the fuel cell. Fig. 5.4 shows various types of fuel cells. Fuel cells have numerous applications. Besides power generation, it can be used as part of a MicroCHP, producing electricity and hot air and water from the waste heat. Furthermore, fuel cells can also be used as a transportation fuel. In fact, many automobile companies are slowly adopting fuel cells and integrating this solution into their fleets. Other advantages of fuel cells include the ability to produce hydrogen in a manner that is environmentally friendly as well as the fact that emissions from a fuel cell are simply water and heat. Moreover, fuel cells have no moving parts, making them more reliable and less susceptible to damage. Lastly, fuel cells have the unique ability to convert chemical energy to electrical energy directly without needing a mediating state. For example, heat engines convert chemical energy into heat, followed by mechanical energy and finally electricity. This shortcut by fuel cells enables them to yield higher exergetic and energetic efficiencies. Moreover, fuel cells enable fuel flexibility and energy security. Diversifying the grid with various sources of energy and obtaining the same commodity from different paths result in more stable and durable energy management as well as a secure energy plan. Fig. 5.5 shows the different types of fuel cells.

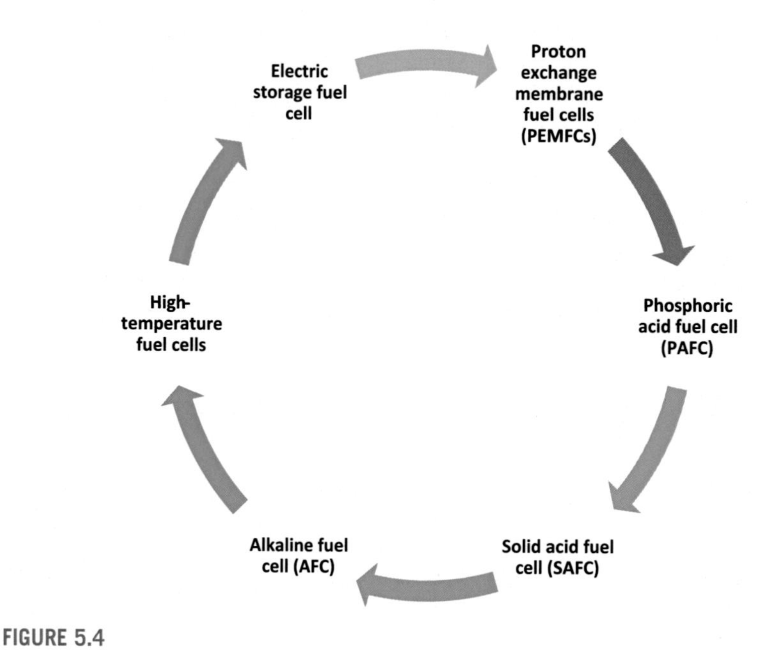

FIGURE 5.4

Types of fuel cells depending on the electrolyte used.

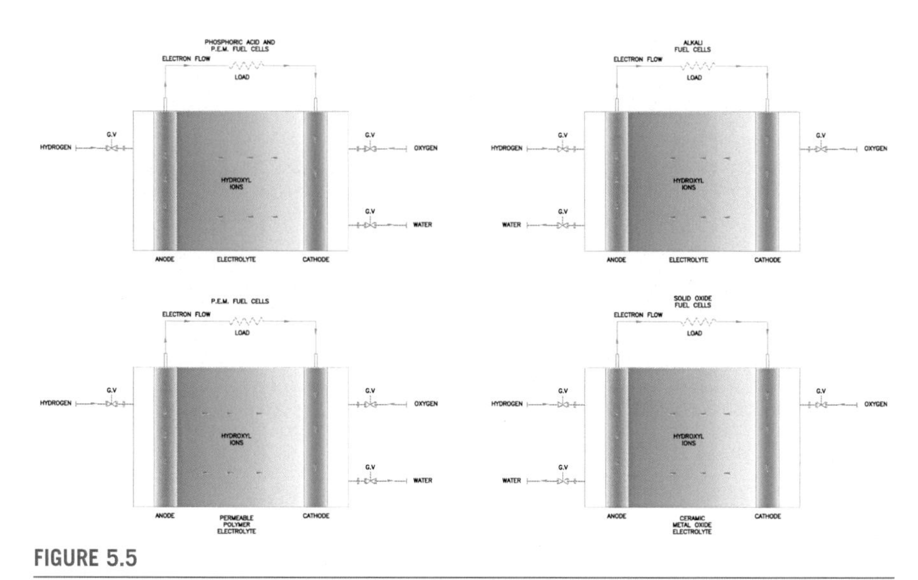

FIGURE 5.5

Structure of fuel cells depending on the type of electrolyte used.

5.4 Photovoltaic thermal energy systems

Photovoltaic thermal (PVT) are hybrid energy systems that are usually localized, small-scale systems that provide electricity and heat for residential or small commercial purposes. It is well known that a rise in cell temperatures causes reduction in the solar systems' overall efficiency. Therefore, PVT systems allow for diverting excess heat away from the cell and utilizing it for heating purposes. As a result, solar cells are cooled, performance is enhanced, and a new commodity is introduced. Fig. 5.6 shows the various types of PVT systems along with a brief description of each.

The cost associated with PVT is definitely higher than the cost of a traditional PV system; however, it would provide two useful output commodities. Furthermore, the heat produced in these systems range between 50 and 65°C, which makes it suitable for space heating. It can also be used as an initial heater for domestic hot water in coordination with another heating source. Factors that impact the PVT performance include the PV cell material, heat transfer fluid or medium, the solar irradiance, the overall system optimization such as the method of integration, and the type of application (Lamnatou and Chemisana, 2017). Moreover, Fig. 5.7 shows the composition of a solar PVT system and its integration within a building envelope.

5.5 Hybrid energy systems

Hybrid power systems can be a combination of systems that are integrated together to generate power. Hybrid reflects the systems' capability to produce and store

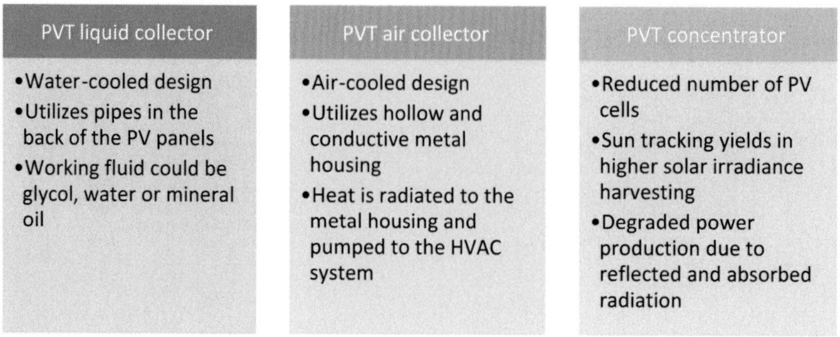

PVT liquid collector
•Water-cooled design
•Utilizes pipes in the back of the PV panels
•Working fluid could be glycol, water or mineral oil
PVT air collector
•Air-cooled design
•Utilizes hollow and conductive metal housing
•Heat is radiated to the metal housing and pumped to the HVAC system
PVT concentrator
•Reduced number of PV cells
•Sun tracking yields in higher solar irradiance harvesting
•Degraded power production due to reflected and absorbed radiation

FIGURE 5.6

Types of photovoltaic thermal solar collectors.

ELECTRICAL GRID
AC
INVERTER
DC
PVT PANEL
SOLAR PREHEAT TANK
HOT WATER TANK
COLD WATER TANK
BLOWER FAN
HOUSE
SOLAR PREHEAT PUMP
PVT PUMP
GSHP PUMP
GHX

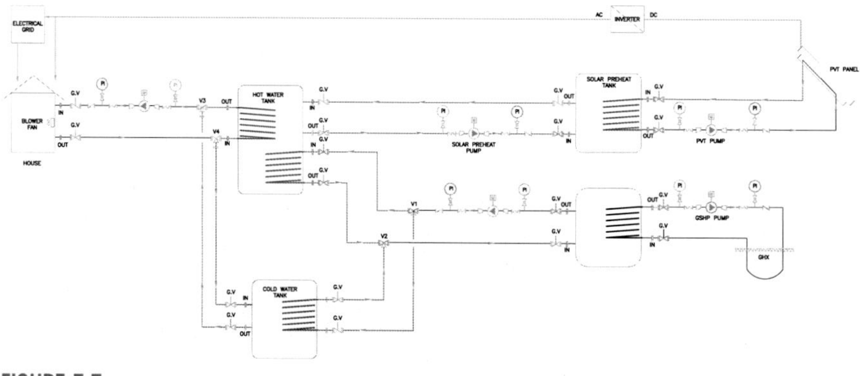

FIGURE 5.7

Illustration for a PVT solar system and its processes.

power. Therefore, these systems involve energy storage of some sort. Furthermore, such systems are strongly linked with renewable energy sources. For example, hybrid systems could involve the combination of solar and wind sources. In addition, a combination of various sources could be achieved to produce the targeted commodity. Hybrid power systems contribute to the energy security and diversity of the energy mix, thus resulting in less power outages. They also provide reliable and durable energy supply. Fig. 5.8 illustrates how two energy sources couple to form a hybrid energy system, which produces energy services.

Although hybrid system designs are popular among renewable energy system designs, it is more practical to integrate hybrid and trigeneration systems with existing fossil fuel-based energy systems. For example, integration of diesel refineries with another renewable energy source for remote communities can provide a more realistic and practical energy hybridization and a more reliable transition to renewable energy. This is because in remote communities, the reliance on fossil fuels such as diesel is absolutely essential. In addition, hybrid systems can also be applied in the

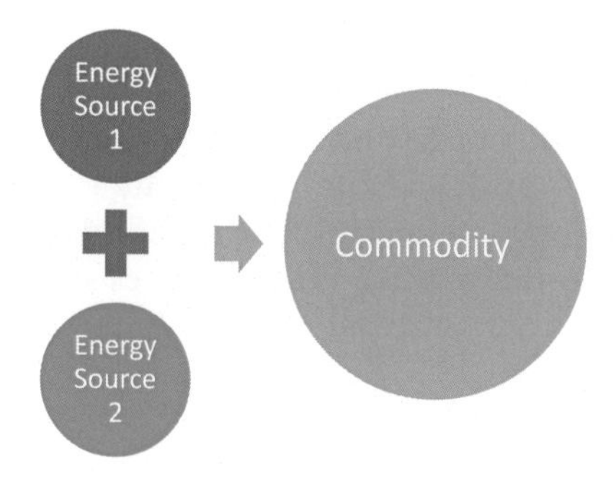

FIGURE 5.8

Illustration of a hybrid energy system.

transportation sector. For example, hybrid vehicles utilize electric power and gasoline as the main sources to power the vehicle. Electric power is preserved using an integrated battery system. Hybrid vehicles lead to lower fuel consumption and consequently better economic performance, thereby leading to more environmentally benign transportation solutions.

Moreover, the diversity of energy supply also causes the battery capacity to be reduced. Similarly, these systems are becoming more popular as stand-alone systems in remote localities. Fig. 5.8 demonstrates various forms of common hybrid systems. Some systems also use diesel as a backup fuel in case one energy source is obstructed. Furthermore, hybrid systems are increasingly becoming a reliable source of power, especially where natural landscape conditions are favorable such as in areas with high solar irradiance, higher and regular wind speeds, and in areas near oceanic ridges, yielding very high underground temperatures. Fig. 5.9 illustrates a sample hybrid system for power generation.

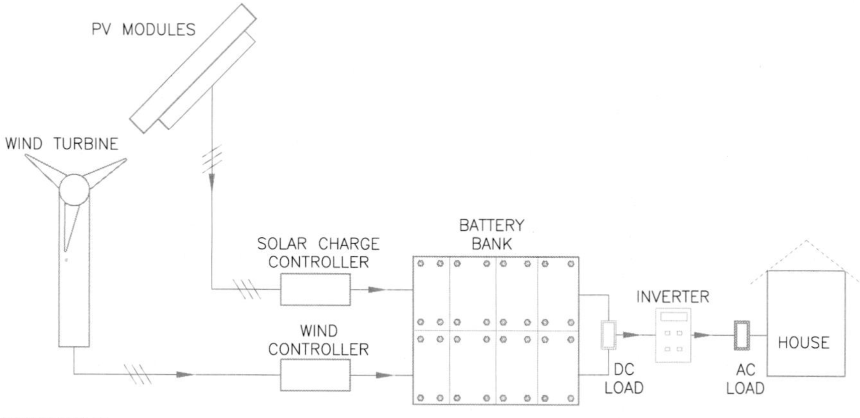

FIGURE 5.9

Sample illustration of hybrid PV/Wind power system for electric generation.

These systems can be designed with flexibility to allow for battery storage, integration with diesel generators, as well as connection to the grid. Therefore, such systems can supply electricity to the grid once meeting the local assigned demand. Grid integration of such systems and sale of electricity varies from jurisdiction to another based on the followed procedures for each jurisdiction. All in all, hybridization of energy systems can be using various sources such as PV/wind sources, PV/diesel, wind/diesel, PV/geothermal, biomass/geothermal, or many other configurations.

5.6 Microgrids

An electric grid is composed of energy generating stations from various sources that feed power to the utility company. In turn, the utility company distributes this energy in an organized fashion to local sectors including residential, industrial, commercial, and institutional sectors. Microgrids are similar to conventional grids; they also use various sources to generate power. However, microgrids are more localized and can remain functional with or without the larger conventional grid. Types of microgrids include campus or institutional microgrids, where the focus is to aggregate existing on-site generation with multiple loads located in tight geography in which their management is central and easy. Another type includes community microgrids that can serve a few up to thousands of customers and support the penetration of local energy (electricity, heating, and cooling). "In a community microgrid, some houses may have some renewable sources that can supply their demand as well as that of their neighbors within the same community. The community microgrid may also have a centralized or several distributed energy storages. Such microgrids can be in the form of an ac and dc microgrid coupled together through a bi-directional power electronic converter" (Chandrasena et al., 2014). Other types of microgrids include remote off-grid microgrids, military microgrids, and finally commercial and industrial microgrids.

Furthermore, physical circumstances such as power outages or economic circumstances have no effect with the presence of microgrids. Therefore, they act as a safety net and a secure energy management option for localities. In fact, microgrids become very useful in events of natural disasters and power outages, as they keep essential institutions such as hospitals and main community hubs revived with electricity. Along these tremendous benefits of cheap, reliable and clean power generation, microgrids also come with challenges. The main challenges pertaining to microgrids include bidirectional power flows control and stability issues. Fig. 5.10 shows various types of applications for a microgrid.

Essentially, a microgrid comprises a local generation source. This source would provide electricity, heating, or cooling, depending on the type of commodity being serviced. Energy storage is becoming an essential component in microgrids as it

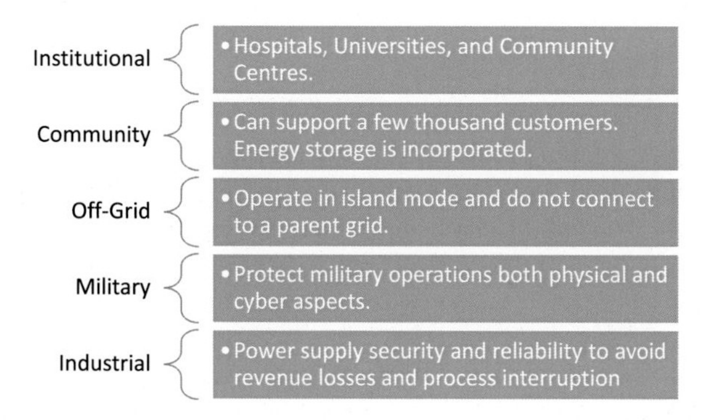

FIGURE 5.10

Various types of applications for microgrids and description for each application.

provides numerous benefits such as increasing the quality of energy, and acting as backup source of energy in emergencies. Microgrids that are connected to a larger grid have a specific point of collection also known as point of common coupling. Fig. 5.11 illustrates a schematic description of a microgrid.

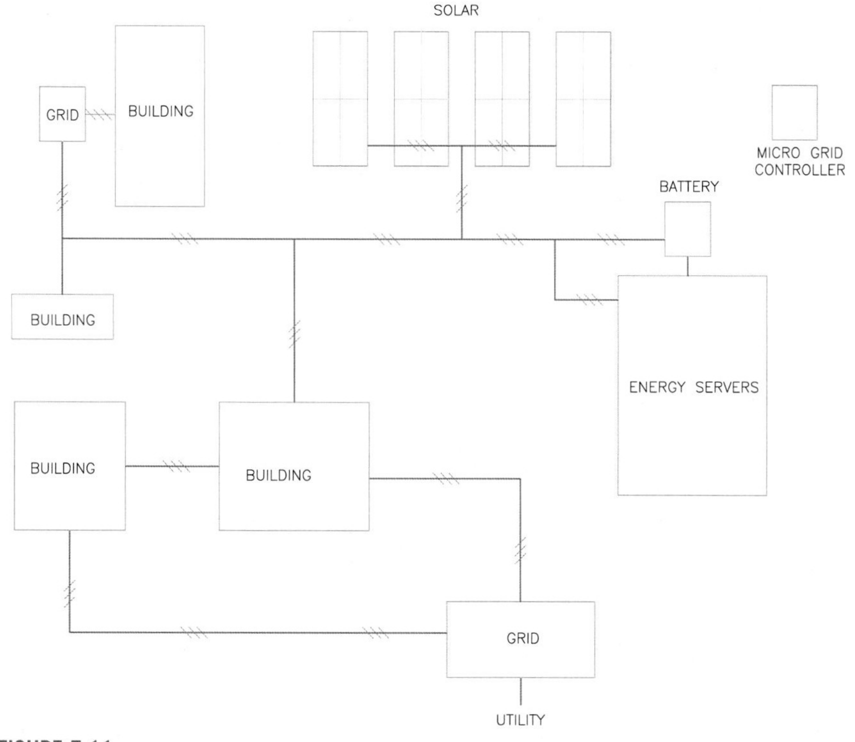

FIGURE 5.11

Schematic of a sample microgrid for a small town and its grid connectivity.

5.7 District heating systems

District heating is another community energy system that provides heat and domestic hot water at a community scale. This system comprises a centralized heating station that distributes heat to a community via insulated pipes. In essence, a larger central boiler acts as the main source of heat as opposed to individual boilers. There are various heat source options for the energy center such as gas combined heat and power energy from waster or renewable energy sources such as heat pumps, biomass, and anaerobic digestion. Hot water or steam travels through an underground network of insulated pipes to a heat interface units located in each building in the case of multioccupancy buildings. Heat is also delivered to commercial buildings and domestic housing. Each house is independent and is able to control their temperature requirements. Heat is delivered through radiators or underfloor heating. Simply, heat is treated like water and electricity and is delivered to the target location. Cooled water or steam returns back to the central boiler to be reheated and circulated again. This type of system is advantageous in many ways including achieving higher energy efficiencies and better environmental friendliness.

Furthermore, the model for district heating is very simple and the degree of ease involved is unmatched. In fact, communities would have no concerns associated with heating repairs or services. Moreover, this system can operate simultaneously with other heat sources. On the other hand, the only risk associated with this system is the quality of insulation for the underground pipes. In the event of leakage or poor insulation, the model will not be financially feasible. Moreover, district heating

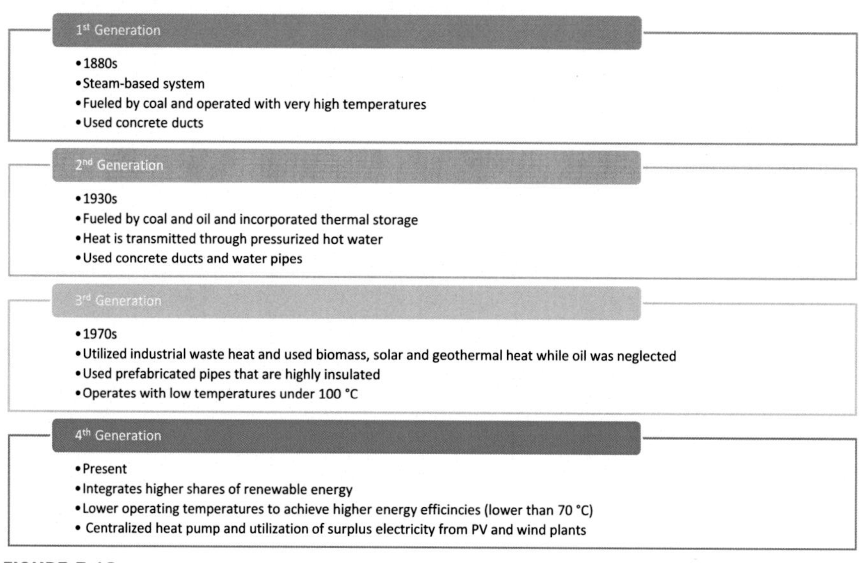

FIGURE 5.12

List and description of the various generation types of district heating systems.

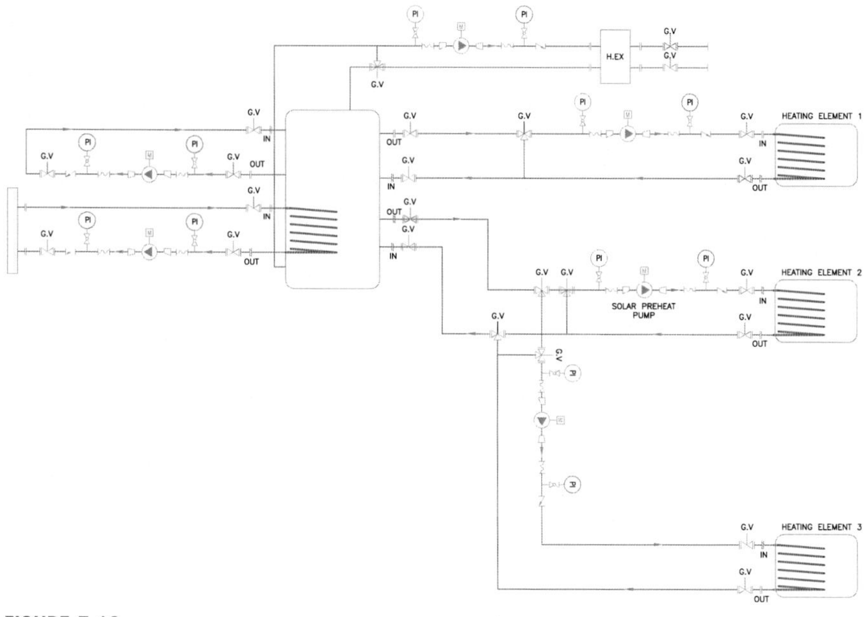

FIGURE 5.13

District heating system utilizing thermal solar and fossil fuels to provide domestic hot water (DHW) and heating.

requires that the serviced homes or buildings be close to one another and not be too scattered. Finally, a significant capital investment is required for such a system. There are four distinct generation types for district heating that developed consecutively since the 1800s. Fig. 5.12 lists and describes each generation type.

Heat generation for district heating systems is achieved through a variety of heating systems including CHP plants, nuclear power plants, and simple combustion power plants that utilize fossil fuels. In addition, renewable energy sources such as geothermal heat, solar thermal systems, and industrial heating pumps that extract heat from waste industrial processes, seawater, or other heat sources. To increase efficiency and reliability of district heating systems, thermal storage units are also incorporated in the design. Thermal storage units can be seasonal, where summer solar heat is collected for redistribution in winter months for instance. A sample district heating system is illustrated in Fig. 5.13 with all components and features.

5.8 District cooling systems

Similar to district heating systems, district cooling systems comprise of a network of insulated underground pipes that deliver chilled water to various users. A centralized production of chilled water is driven by renewables, compressor-based chillers, absorption chillers, or other sources such as deep lake cooling. In fact, cooling is treated as a deliverable commodity exactly like water and electricity. This system saves buildings from having their own individual chillers, air conditioners, boilers, and furnaces. Therefore, district heating systems allow for significant savings from capital and operating costs associated with the equipment abovementioned. Furthermore, district cooling systems provide more reliability, environmental sustainability, and flexibility. Moreover, this technology offers highly efficient processes leading up to more than 40% of energy efficiency improvements and 20% lifecycle cost savings compared to conventional air conditioning systems. In addition to capital and maintenance savings, district cooling systems decrease the demand on the grid. Conventional cooling systems usually consume 20%–35% more electricity, which is required during peak hours, accumulating significant costs. However, with the incorporation of district cooling systems, substantial electricity costs are all avoided, causing less pressure on the grid and better energy security.

In fact, traditional air conditioning systems create 50%–70% of the peak electricity demand in a building, usually at peak cost as well. Therefore, with district cooling systems, peak power demands on the grid are avoided and costs are reduced. Moreover, the distribution network of pipes is designed by creating a pressure difference between supply and return lines. Delivered chilled water can either interphase with heat exchangers at each building or can directly be delivered to the buildings fan coils. Similar to district heating, district cooling systems require that user buildings be close to each other and thus such systems are most favorable for dense cities or institutions with numerous buildings such as university campuses. The aggregation of the cooling demand for multiple buildings, district cooling

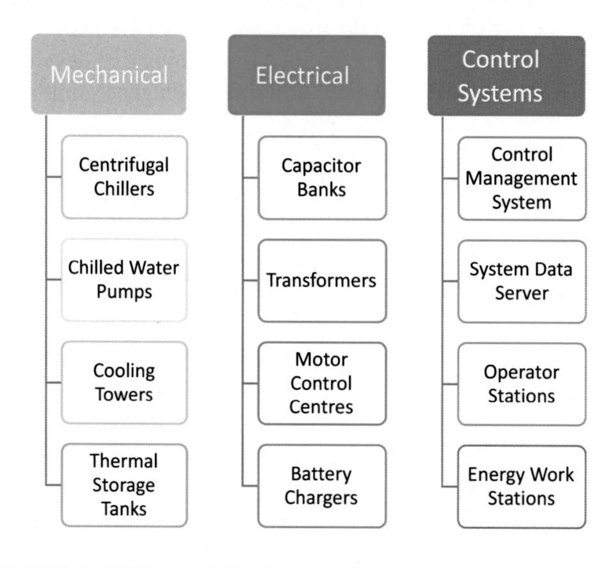

FIGURE 5.14

District cooling system components including mechanical, electrical, and control systems.

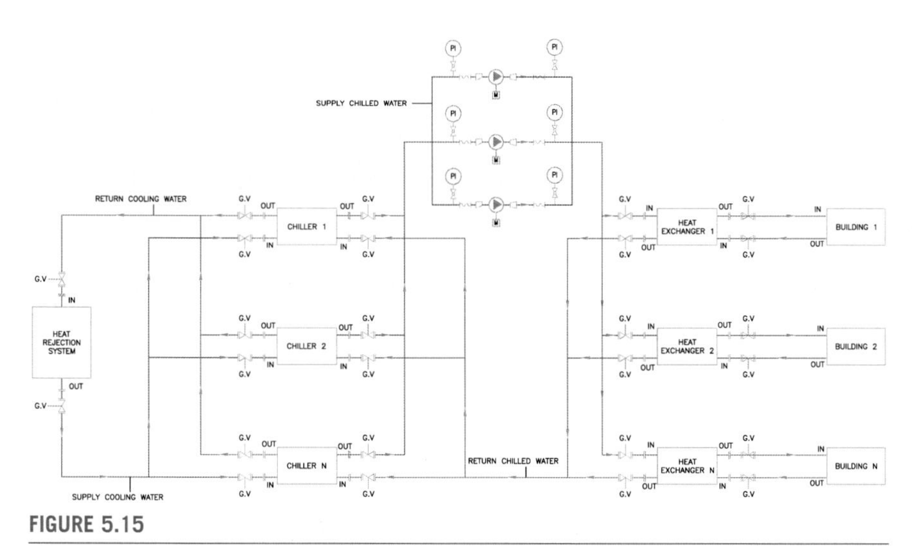

FIGURE 5.15

District cooling system using deep water as cooling source.

systems prove financial competence and better environmental friendliness. Fig. 5.14 shows the different components required in a district cooling plant.

District cooling systems are advantageous, as they are environmentally friendly, yield lower lifecycle costs, are reliable, decrease building costs, and provide architectural flexibility. They have the ability to use alternative and renewable energy sources, resulting in more than 50% reduction in power consumption. They do not need any refrigerants, reduce greenhouse gas (GHG) emissions, and provide higher flexibility for cooling loads. As for maintenance of machines, they decrease the construction of individual cooling systems. Furthermore, district cooling systems present a reliable cooling source, as they are unpretentious by peak load demands and grid interruptions. Fig. 5.15 shows a sample district cooling system.

5.9 Thermal energy storage

Thermal energy storage (TES) is a technology that enables thermal heat storage to be stored for later use. Depending on the scale of the storage, heat can be stored for hours, days, months, or seasons. Furthermore, this heat stored can be used for heating, cooling, or power generation applications. They are particularly useful for larger buildings and industrial sites. In essence, TES acts as a big battery storage that can be utilized in various ways. Furthermore, TES systems prove better financial and environmental performance by increasing overall efficiency, reducing GHG emissions, and reducing costs. Efficient TES systems use a variety of materials with varying thermos-physical properties to allow for better storage efficiencies. Moreover, TES systems are closely associated with solar systems. The intermittency and the inconstant availability of solar radiation make the integration of TES systems ideal.

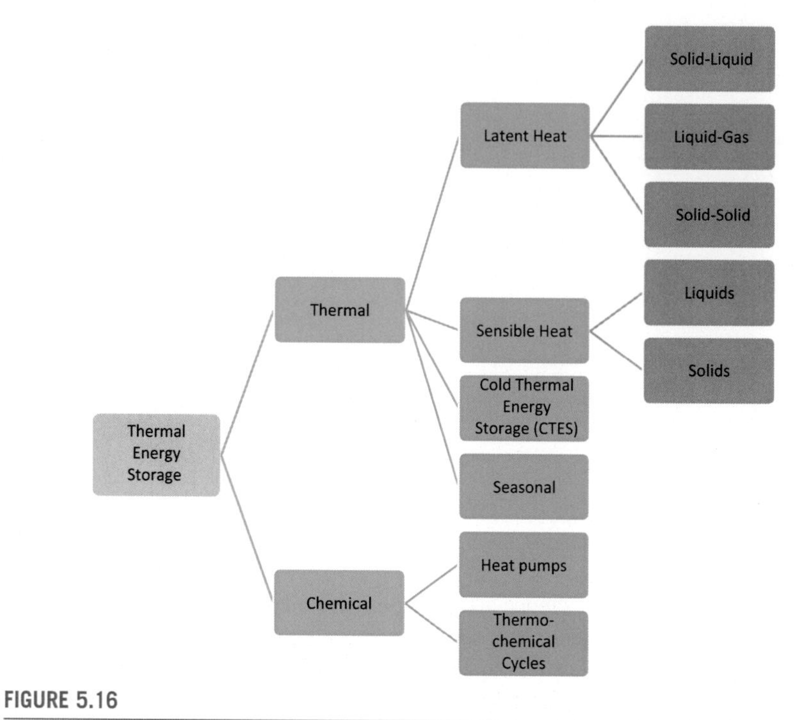

FIGURE 5.16

Types of thermal energy storage and the types of heat used.

Modified from Sarbu and Sebarchievici (2018)

In fact, TES empowers the solar technology to become more reliable and effective to various applications. There are various types of TES depending on the method for heat generation. Fig. 5.16 shows the different types.

Thermal energy storage can be achieved by lowering the temperature of a substance or changing its phase or by combining both together, resulting in thermal storage. TES involves three main steps: charging, storing, and discharging. The storage medium varies depending on the type of TES and intended application. For instance, water is used as the medium in sensible heat storage because of its lucrative attribute of having one of the highest specific heats of any liquid at ambient temperatures. Being in the liquid form makes it easier for water to be transported thermally and allows for better heat transfer rates. Moreover, TES has been a research interest in mechanical engineering, specifically the heat and mass transfer, and transient behavior performance. Moreover, PCMs have been explored for their use in energy conservation in buildings, and thermal analyses have been conducted. In addition, the use of different storage materials and the use of aquifer TES have been investigated in depth in addition to the feasibility of solar ponds and water TES. Thermal energy storage systems introduce a vast number of benefits such as increasing the generation capacity to meet the growing demand for heating, cooling, and power. Furthermore, stored energy can be dispatched for use during peak times for

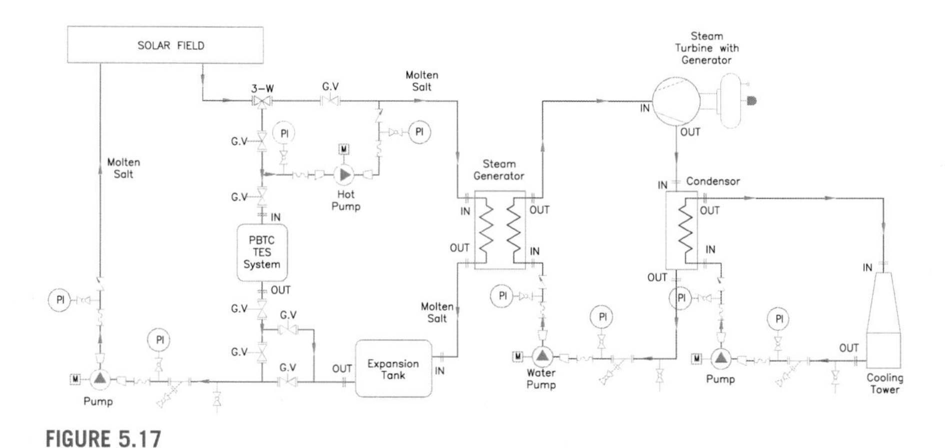

FIGURE 5.17

Thermal energy storage system using concentrated solar power.

electricity, yielding significant savings by avoiding electric use from the grid in peak times. TES can be evaluated following a set of criteria including technical, environmental, and economic criteria. For example, technical criteria that need to be considered include storage capacity, size, cost, lifetime, efficiency, resource use, commercial viability, and safety. On the other hand, environmental criteria include the system design, emissions, material, and operational practices used for the TES as well as the various processes throughout the lifecycle of the TES.

Furthermore, economic criteria include hourly thermal loads for the peak day, comparison between the electrical load profiles of a conventional system versus a TES system, as well as the size of the storage system (Dincer and Rosen, 2011). Fig. 5.17 demonstrates a sample thermal storage system.

5.10 Cogeneration systems

Cogeneration systems refer to energy systems that have the ability to produce two useful commodities simultaneously. A good example of cogeneration systems are combined heat and power plants, where electricity and useful heat are both produced from one plant. Cogeneration is a highly efficient energy orientation that can achieve primary energy savings compared to conventional power and heat supply. Furthermore, aside from electricity and heat combination, commodities include heat for fruit drying, desalination, and chemical applications. Therefore, cogeneration applications include residential, commercial, and industrial applications. In essence, cogeneration plant is very similar to the CHP systems explained earlier. Moreover, 11% of the total electric generation in European Union is due to cogeneration plants. In fact, the world's most intensive cogeneration economies include Denmark, the Netherlands, and Finland. Furthermore, 82% of the Finnish electricity was produced using cogeneration plants by conventional thermal power plants. Fig. 5.18 shows a

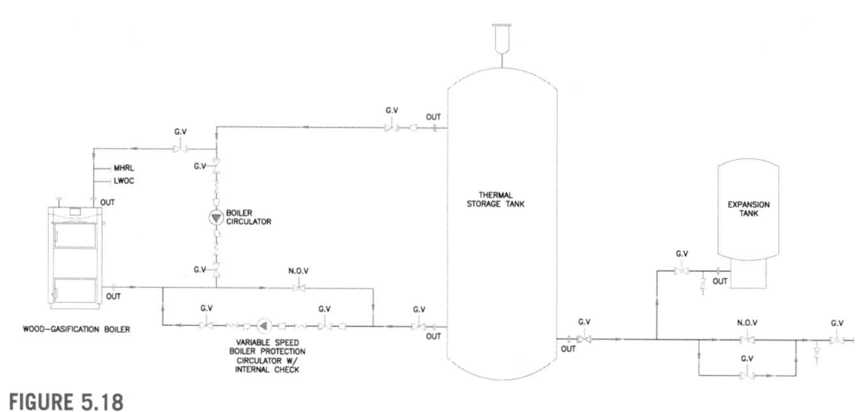

FIGURE 5.18

Biomass-powered cogeneration power plant to provide electricity and district heating.

sample cogeneration power plant. Conventional power plants such as coal, microturbine, natural gas, nuclear power, oil, or small gas turbine can be converted to combined cooling, heat, and power system plants. Thermal efficiency in a cogeneration system is defined as follows:

$$\eta_{th} = \frac{W_{out}}{Q_{in}} \tag{5.1}$$

where η_{th} denotes the thermal efficiency, W_{out} represents the electric power output in addition to the heat output, and Q_{in} is the total heat input to the system.

Cogeneration systems denote a very favorable energy solution for communities and districts, as it brings a vast variety of benefits such as increase system efficiency. In fact, it is the most effective and efficient method for power generation. Furthermore, cogeneration limits the GHG emissions very successfully and enhances processes that lead to substantial cost savings and affordable electricity rates. Moreover, such systems pose an opportunity to increase decentralized electricity generation, thus avoiding transmission losses, and increasing system flexibility. In addition, local energy security is enriched by increasing the diversity of energy supply and ensuring reliable energy supply.

5.11 Trigeneration systems

As the name denotes, trigeneration systems provide three useful commodities simultaneously from one process. In addition to cogeneration systems, a chiller is integrated to provide cooling. To achieve this, waste heat is converted to chilled water through absorption systems. Therefore, trigeneration systems achieve even higher efficiencies, as they utilize the waste heat from the cogeneration systems to provide a useful commodity. Fig. 5.19 illustrates how trigeneration systems function. Heating can be used to provide space heating, domestic hot water, trigger

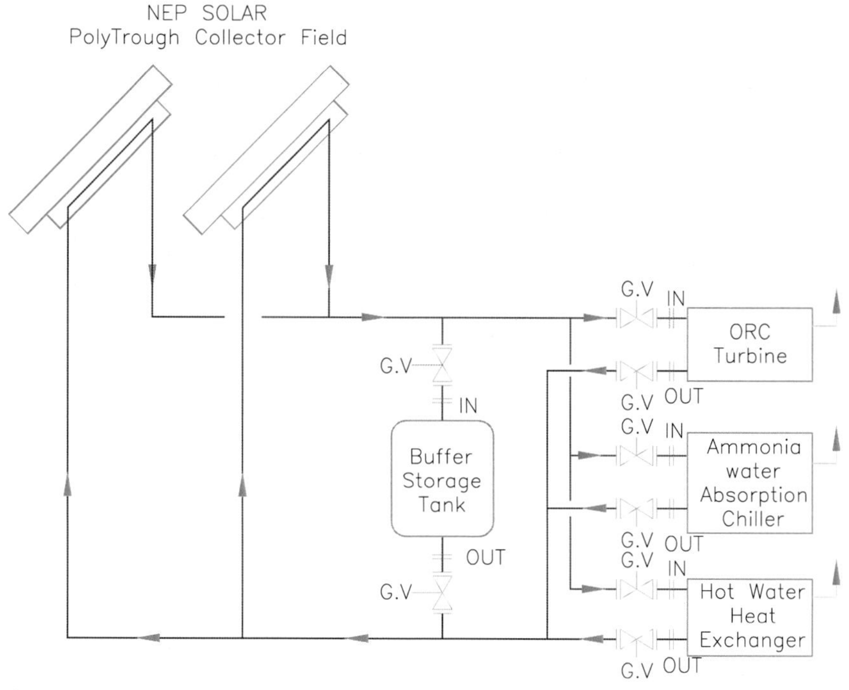

FIGURE 5.19

Solar-powered trigeneration power plant to provide electricity, DHW, and cooling.

industrial or chemical processes, or can be stored using thermal energy storage. Similarly, cooling can be used to provide HVAC solutions to buildings, commercial or industrial sites. Furthermore, electricity is used to power residential, commercial, institutional, and industrial sectors. Moreover, the source of energy for a trigeneration system can vary as discussed earlier in previous chapters. They could use fossil-based fuels, solar energy, geothermal energy, biomass, nuclear, or other sources of energy mentioned earlier.

5.12 Closing remarks

In conclusion, this chapter summarized various types of energy systems that are used for communities or smart cities. Designs for these systems have been thoroughly presented and investigated in this chapter. Moreover, these systems feature an integrated approach and sustainable solutions for communities to provide main necessary commodities such as heating, cooling, and electricity. Moreover, as communities evolve, the energy demand tends to increase and thus various energy systems and models have been developing to attend to these necessities.

Furthermore, the integration of systems helps to preserve energy and exergy efficiencies and decrease exergy destruction and losses. In addition, thermal energy storage solutions increase the viability and efficiencies of energy systems, especially when dealing with district heating and cooling. Cogeneration and trigeneration systems result in higher system efficiencies and less exergetic destructions and losses. Hybridization of energy systems can occur with existing fossil fuel-based energy systems or other renewable energy systems. In closing, smart cities require detailed planning when it comes to energy systems, where digital solutions are integrated and utilized for better energy performance as well as waste heat recovery.

CHAPTER

Sustainability modeling

6

6.1 Introduction

Sustainability is a very complex concept with multidisciplinary relationships. As presented in the previous chapter, there is no exact definition for sustainability; however, the Brundtland Commission of the United Nations' definition on sustainability is the most popular one. Although researchers and scientists have proposed many sustainability assessment models, until today there is not a universally recognized assessment model. This opportunity to formulate an integrated sustainability assessment model, which is both robust and simple, is extremely valuable. There have been many attempts and a number of sustainability assessment methodologies, which exist for evaluating the performance of companies, cities, energy systems, and product manufacturing (Ramachandran, 2000). In fact, sustainability management in industries was triggered by the evolution of a number of sustainability literature such as the World Business Council for Sustainable Development (WBCSD, 1997), the development of sustainability standards (OECD, 2004), and the emerging of the Global Reporting Initiative (GRI, 2002a,b). Initiated by the United Nation Environment Programme in partnership with American environmental organizations, the GRI was launched in 1997 with the intent of enhancing the quality, structure, and layout of sustainability reporting. On that note, the UN Commission on Sustainable Development (CSD) introduced a list of 140 indicators that touch on various aspects of sustainability (CSD, 2001). Although the GRI assesses sustainability based on three major categories (social, economic, and environmental), the Wuppertal Institute developed a sustainability framework, which examines the four categories identified by the United Nations CSD (social, economic, environmental, and institutional). Some gaps are evident in the development of sustainability assessment that is noted after reviewing the literature. These include, but are not limited to the following:

- Absence of universally adopted and shared understanding of sustainability.
- Sustainability assessment models that are specifically geared toward a specific application.
- The focus on specific categories and neglect of major categories in some models.

Energy Sustainability. https://doi.org/10.1016/B978-0-12-819556-7.00006-1

- The double-counting trend when investigating similar indicators.
- The absence of target/reference values to compare to actual values. Life cycle values are given a dimensionless score without relationship to a target value.

6.2 Sustainability assessment categories

Because sustainability is a multidisciplinary concept, various domains must be investigated and explored to achieve a comprehensive assessment. The following section elaborates more on the various impact domains.

6.2.1 Energy aspect

Energy analysis is a thermodynamic tool that has been used to assess energy systems' sustainability. Thermodynamic-based indicators in sustainability are novel, and the concept is slowly growing within the scientific community. Utilizing the first law of thermodynamics and assessing sustainability quantitatively based on accurate and robust thermodynamic data is an added advantage. Gnanapragasam et al. (2010) introduced a methodology for assessing the sustainability of hydrogen production from solid fuels. This methodology incorporated different aspects of energy under different categories in their research. Energy rate, denoting the rate at which energy can be supplied by the element or process, was assessed along with net energy consumption, which is referred to the energy requirement of the element to transport it to the point of use and utilize it in the operation of processes. Moreover, efficiency was also a key factor in the sustainability assessment of this model. Caliskan et al. (2011) investigated a solar ground–based heat pump with thermal energy storage. The sustainability of this model included energy analysis comprising energy input rate (solar radiation), energy storage rate, efficiency assessment derived from the collector, and other heat losses. More recently, alternative fuels were assessed for Thailand to replace the current diesel fuel, which is widely used. In this assessment, the net energy ratio is used as an indicator within the life cycle assessment (LCA) of the proposed alternatives (Permpool and Gheewala, 2017). Furthermore, Lo Piano and Mayumi (2017) developed an integrated assessment model using multiscale integrated analysis of societal and ecosystem metabolism (MuSIASEM) to investigate the performance of photovoltaic power stations for electricity generation. Economic indicators with reference to energy such as energy payback time and energy return on investment have been suggested. These indicators will be discussed under other categories. In this study, a major category was energy accessibility. This factor was measured by considering the primary energy source type and energy carriers. Caliskan et al. (2013) modeled hybrid energy systems namely hybrid geothermal energy–wind turbine–solar photovoltaic panel, inverter, electrolyzer, hydrogen storage system, and proton exchange membrane fuel cell. Energy analysis was a pillar tool that they used to model these systems. Energy analysis included energy balance, the net energy input rate, the energy loss in the form of heat, the electric power production, and the electricity production. Furthermore, other proposed sustainability measurement approaches of renewable energy productions include that of Pierie et al. (2016) when they focused on modeling the green gas production pathways. In this model, they measured direct and indirect energy flow rates to determine efficiency. Moreover, Dincer and Zamfirescu (2012) introduced a new thermodynamic concept to greenize energy systems. For this, they assess the energy efficiency and energy balance and mass equations of their case studies. In addition, the environmental impact of energy systems is associated with the energy-resource utilization along with inexpensive and stable energy supply (Dincer and Rosen, 2011). Hydrogen fuel cell systems were examined by Dincer and Rosen (2011) for sustainability. Energy was a major aspect of determining the sustainability as they measured the efficiency of the hydrogen fuel cell systems along with considering other aspects. The summary of these energy indicators is illustrated in Fig. 6.1. In summary, it is evident that energy analysis in various forms, stages, and capacities is incorporated in sustainability assessment since the last 2 decades. The use of thermodynamic-based variables in presenting quantifiable values that mirror the sustainability of energy systems is of utmost importance. Thus, in this book, the energy efficiency and the production rate (TWh/year) have been selected to be the factors that will simply and accurately assess the sustainability of energy systems based on energy performance. As can be observed from the literature, the reasons behind

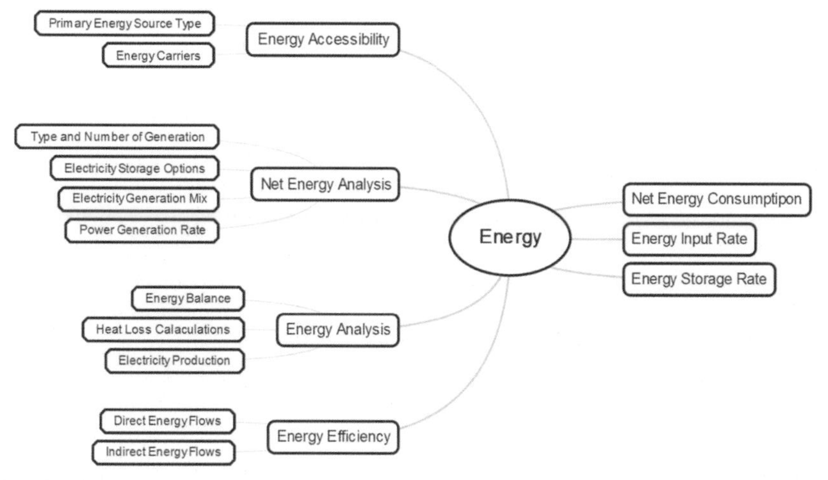

FIGURE 6.1

Concept map of the energy category and the distribution of various parameters used in the literature to account for the energy factor in sustainability assessment.

the variety of energy indicators presented in the mind map lie in two main reasons: first, studies varied in their nomenclature of different indicators and most indicators could be grouped accordingly to avoid allocation of results. Second, studies differed on their research focuses and thus certain parameters were selected based on the specific system being studied.

Overall, it is suggested that the efficiency of the energy system defined as the ratio of the useful energy output given by the energy system to the total energy expended is a comprehensive parameter that will accurately reflect the sustainability aspects of the energy analysis of systems. The production rate is also suggested as systems that have higher energy productions than others with competitive energy demands are also more beneficial and more attractive than systems that have high-energy demands with low energy output.

6.2.2 Exergy aspect

Exergy analysis is another thermodynamic-based tool that has been emerging into the sustainability assessment models. It is related directly to the second law of thermodynamics, where exergy is always destroyed when a system involves irreversible reactions such as heat loss to the environment. Another factor that emerges is the entropy of the system in relation to the destruction of exergy. Therefore, exergy analysis is a vital tool that can be used specifically to assess energy systems' sustainability (Dincer and Rosen, 2007). Those systems with higher exergy efficiencies reflect more sustainable processes that do not have much irreversibility and vice versa. Dincer (2007) conducted sustainability assessment on green energy systems and concluded that the use of exergy analysis as a tool for assessment of green energy systems is essential to increasing efficiency, and decreasing environmental effect. In fact, in 2004, exergy was highlighted as a driver for achieving sustainability by Dincer and Rosen (Dincer and Rosen, 2004). Exergy is a measure of deviations between the system and environmental equilibrium. They concluded that the potential usefulness of exergy analysis in addressing sustainability matters and resolving environmental challenges is substantial. Hacatoglu et al. (2016) studied the sustainability of a wind–hydrogen energy system and applied their novel assessment index. Exergy efficiency was measured by finding the ratio between exergy outputs and the exergy of the wind. When compared to the gas-fired system, the exergy efficiency considered the chemical exergy of the fuel as part of the calculation. Exergy destruction ratio has been used as an indicator in the sustainability assessment of hydrogen production from solid fuels by Gnanapragasam et al. (2010). This tool was utilized to assess the technological aspect of the sustainability assessment of this study.

Furthermore, Caliskan et al. (2011) particularly selected exergy analysis as a function of the sustainability assessment of the solar ground–based heat pump with thermal energy storage. In this study, the rate of maximum exergy input, exergy losses, and exergy storage have been determined based on reference parameters. Exergy efficiency was also considered in this study. Moreover, even more exergetic indicators have been introduced and proposed by other scientists in the literature. For example, Midilli et al. (2012) researched about the environmental and sustainability aspects of a recirculating aquaculture system. In their study, they proposed the following exergetic indicators: exergetic efficiency, waste exergy ratio, exergy recoverability ratio, exergy destruction ratio, environmental impact factor, and exergetic sustainable index. Caliskan et al. (2013) conducted exergy efficiency analysis on the case studies they performed as an indicator of sustainability. Caliskan et al. (2011) also considered exergetic parameters when assessing the sustainability of three types of air-cooling systems for building application. They namely used the specific exergy flow, exergy efficiency, and specific exergy destruction as indicators of the exergetic index of these air-cooling systems. Furthermore, Caliskan et al. (2012) conducted exergeoeconomic and sustainability analyses on a novel air cooler in which they used exergy input, output, loss, destruction rates, exergetic coefficient of performance, primary exergy ratio, and exergy efficiency. On the economic side, they also calculated the exergetic cost rate. Dincer and Acar (2016) highlighted the importance of investigating irreversibilities, energy, and exergy efficiencies and considered them as critical steps to obtain sound sustainability assessment. The summary of these exergy indicators is illustrated in Fig. 6.2.

As exergetic analysis is highlighting its importance in determining sustainability of energy systems, its use is growing steadily. Exergy analysis has not been as commonly used as energy analysis, yet it is equally important as it relates to the second law of thermodynamics. Exergy adds to the quantitative parameters that are used to assess sustainability. Such parameters shape the sustainability model and add robustness and accuracy to the model by involving data-based factors in the assessment and limiting the subjectivity of the assessment as much as possible. Exergy as presented in Fig. 6.2 has been assessed through a number of indicators and parameters. Efficiency has been redundant throughout many studies while other factors such as storage, losses, and exergy destruction have been specifically selected to measure aspects of the system that otherwise would be unknown. There is some

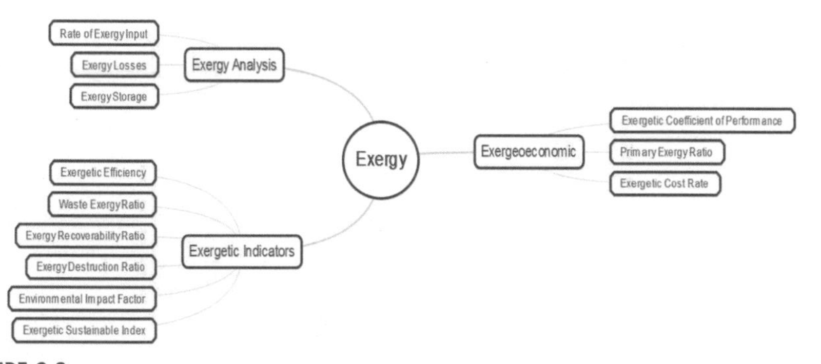

FIGURE 6.2

Concept map of the exergy category and the distribution of various parameters used in the literature to account for the exergetic factor in sustainability assessment.

overlap between the exergy indicators and therefore, the exergy analysis is limited to two main factors. In this book, efficiency (percentage) and exergy destruction ratio are selected as the main indicators that will adequately and accurately reflect the exergy performance of the target energy systems. From the second law of thermodynamics, it is established that no system will ever be 100% efficient. Thus, exergy efficiency is a reliable tool that has been repetitively used in the literature to measure the exergy of systems. As all thermodynamic processes are governed by the laws of conservation of energy and mass, the laws entail that the mass and energy can be neither destroyed nor created in a process. As exergy is not conserved, it is therefore destroyed by irreversible processes within the system. Therefore, exergy destruction is an integral part of the exergy balance calculation. It is proposed that exergy efficiency and the exergy destruction ratio are two indicators that will comprehensively echo the exergy of the energy system of interest. In fact, using the exergy as a tool to assess sustainability yields performance improvements, efficient analysis, and effective design of energy systems (Dincer, 2007). Other variables would not be necessary in this model to avoid replication of results and thus exaggerating the effect. Overall, the objective behind this category is to deeply understand the efficiency and performance of the energy system thermodynamically.

6.2.3 Economic impact

Economic aspects have always been in the core of sustainability assessment along with the environmental and social factors. Understanding the financial repercussions and outcomes of projects or energy systems is vital to understanding its sustainability. Furthermore, the economic categories are very crucial to decision-making in various industries as well as to government agencies. Therefore, various sustainability assessment studies across the literature investigated the economic aspect to reach a reasonable sustainability assessment model (Fig. 6.3). Braganca et al. (2010) took

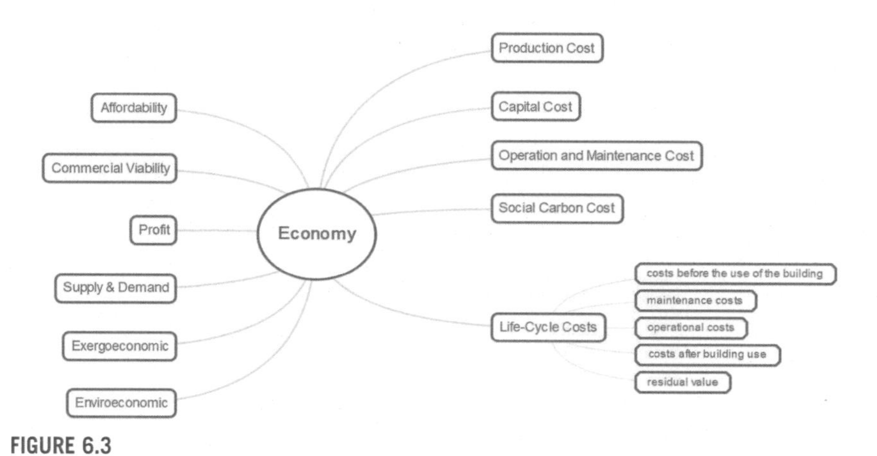

FIGURE 6.3

Concept map of the economic category and the distribution of various parameters used in the literature to account for the economic factor in sustainability assessment.

into account the economic performance when assessing building sustainability for instance. Namely, they considered life cycle costs including costs before the use of the building, maintenance costs, operational costs, and costs after building use and residual value. Furthermore, Hacatoglu (2016) considered the affordability and commercial viability of energy systems to assess their economic effect. Economic assessment is also present in the works of Jin et al. (2019), Zhang et al. (2012), and Balfaqih et al. (2017). On the other hand, Gnanapragasam et al. (2010) considered some economic factors, yet they were dispersed as indicators of sociological and technological categories. They considered economic benefit, policy, and per capita demand. Dincer and Acar (2016) considered a number of indicators in their design including production cost, investment cost, operation and maintenance cost, and social carbon cost. The exergeoeconomic and enviroeconomic aspects were examined for a novel air cooler (Caliskan et al., 2012). The exergeoeconomic tool is used to assess the exergetic cost, which translates to exergy-based economic analysis, in which costs are distributed among outputs. The method shows correlations between capital costs, working hours, and exergy destruction.

The enviroeconomic tool was mainly composed of the carbon dioxide emission price. Paolotti et al. (2017) examined the economic assessment of agro-energy wood biomass supply chain. Santoyo-Castelazo and Azapagic (2014) included three main indicators to assess the economic category of their sustainability model. They used capital costs, total annualized costs, and levelized costs to assess the sustainability of energy systems. Fig. 6.6 shows the various economic indicators used in the literature to assess the economic category of sustainability assessment.

In this book, some of the indicators found in the literature are used while others are enhanced in a way that the economic impact of the energy system is clearly and concisely factored in the sustainability assessment. Fig. 6.4 shows the indicators proposed for this model. These three elements of the economic category relate to each other, as they are able to affect one another. For example, if an energy system has a

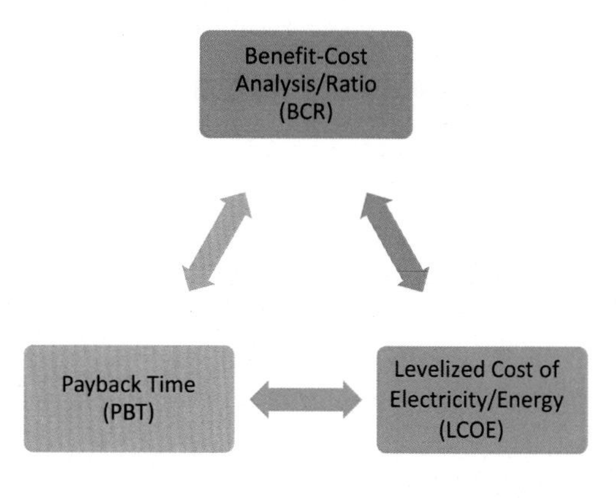

FIGURE 6.4

Economic indicators proposed for the sustainability assessment in this model.

high benefit-cost ratio, the levelized cost of electricity would be lower and the payback time of the system would be shorter. On the contrary, if the payback time of the system is long, the benefit-cost ratio decreases and thus the levelized cost of electricity peaks. The levelized cost of electricity represents the affordability aspect of sustainability.

Affordable energy is definitely more sustainable. Similarly, energy systems must be profitable and economically sound from a business perspective. Thus, the benefit-cost ratio cannot be in the negative and must always maintain a good margin. Maintaining high benefit ensures the speed in paying back the initial capital investment of the system, which enables the system to collect profit thereafter. Therefore, these three concepts of the economic domain are interrelated and influence one another.

6.2.4 Technology

Indicators to reflect the energy system's technology is novel to sustainability assessment. Different from economic, social, and environmental categories, this technology category brings another outlook to the suitability assessment of energy systems. There are not many indicators found in the literature to reflect the technology of the product. Gnanapragasam et al. (2010) had a technological category in his proposed model. This category was assessed along with sociological and ecological categories with 10 indicators for each category. In the technological categories, they mixed between energy, exergy, efficiency, and actual technology-related indicators (Fig. 6.5).

Specifically, they examined demonstration, commercialization, impact, and evolution of the technology as indicators toward the assessment model. Lo Piano and Mayumi (2017) had technological achievability along with three other pillar categories as the basis of their model. Pierie et al. (2016) did not consider the technological aspect in detail; however, they assessed the lifetime of the system. The technological lifetime of a green gas production pathway was taken into account as part of the long-term dynamics of the system. Fig. 6.7 illustrates the different indicators that have been used in previous studies to assess the technology category.

This book expands more on this category to obtain realistic assessment of the technology aspect. Fig. 6.6 shows the indicators proposed for this assessment model.

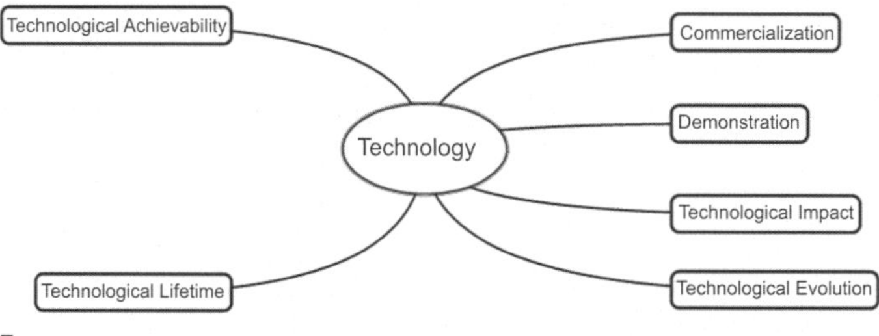

FIGURE 6.5

Concept map of the technology category and the distribution of various parameters used in the literature to account for the technology factor in sustainability assessment.

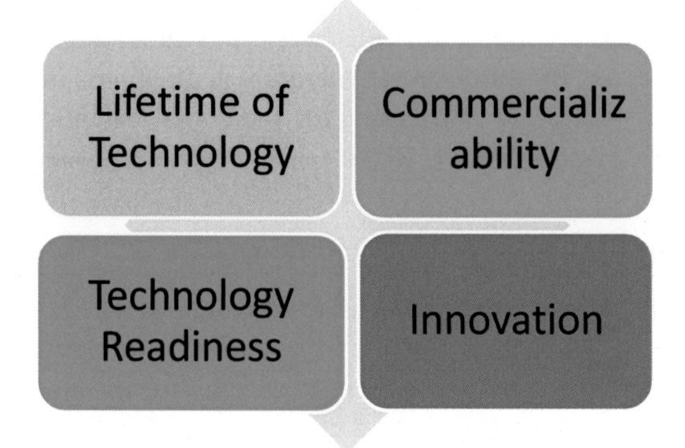

FIGURE 6.6

Technological indicators proposed for the sustainability assessment in this model.

Systems vary in their technological competitiveness. Some are commercially viable with large-scale market competition while other systems are yet to evolve. A system could be technology-ready, but not commercially viable. Furthermore, the lifetime of systems varies depending on many manufacturing, operation, and maintenance factors. Further detailing of each of these indicators will be discussed in the next chapter.

6.2.5 Social aspect

The social category of the sustainability assessment is majorly composed of qualitative and subjective indicators. Social aspects have been an integral part when discussing sustainability assessment models along with environmental and economic categories. In the literature, there have been various social indicators to account for this category in different models. Some of the indicators within this category are categorized as weak indicators while others are considered strong. When assessing building sustainability, Braganca et al. (2010) used various social indicators that related to buildings rather than energy systems. They used indicators such as hydrothermal comfort, indoor air quality, and visual comfort. Gnanapragasam et al. (2010) used 10 social indicators to assess this category. Indicators used included human resources, public opinion, living standards, and human convenience along with other economic and environmental aspects. The social cost of carbon was examined by multiple studies, namely Dincer and Acar (2016). Lo Piano and Mayumi (2017) examined the socioeconomic effect as part of their integrated assessment of the performance of photovoltaic power stations for electricity generation purposes. Afgan and Carvalho (2004) covered the social category by focusing on job and area indicators. Job indicator represents the number of hours of new job to be opened corresponding to the respective option in the following 10 years. The area indicator represents a parameter, which defines the number of meter squared per unit power. Santoyo-Castelazo and Azapagic (2014) investigated the social category more comprehensively than other studies by investigating security and diversity of supply of energy, public acceptability, health and safety, and intergenerational issues.

The summary of used indicators in the literature is illustrated in Fig. 6.7. Overall, the proposed list of indicators to account for the social category is illustrated in Fig. 6.8.

6.2.6 Environmental impact

Sustainability has been associated with environmental aspect since its inception. In fact, the harmful environmental impacts of conventional energy sources, which triggered global warming and climate change account for the emerging of modern sustainability. Thus, the literature is filled with sustainability assessments, models, and reviews encompassing environmental factors. Environmental factors have been even considered in building sustainability assessments. Environmental performance included climate change, emissions to air, water and soil, water efficiency, and resource depletion (Braganca et al., 2010). Most studies in the literature use the LCA to measure the environmental impacts of the systems. Braganca et al. (2010) considered local environmental impacts along with cultural aspects in their research. Although they highlighted energy as a major key issue when addressing sustainability, they did not use it in their model. Hacatoglu (2016) considered more environmental factors in his assessment. In fact, three major assessment categories were environmentally based while the other three accounted for technical, economic, and efficiency parameters. He used 12 environmental

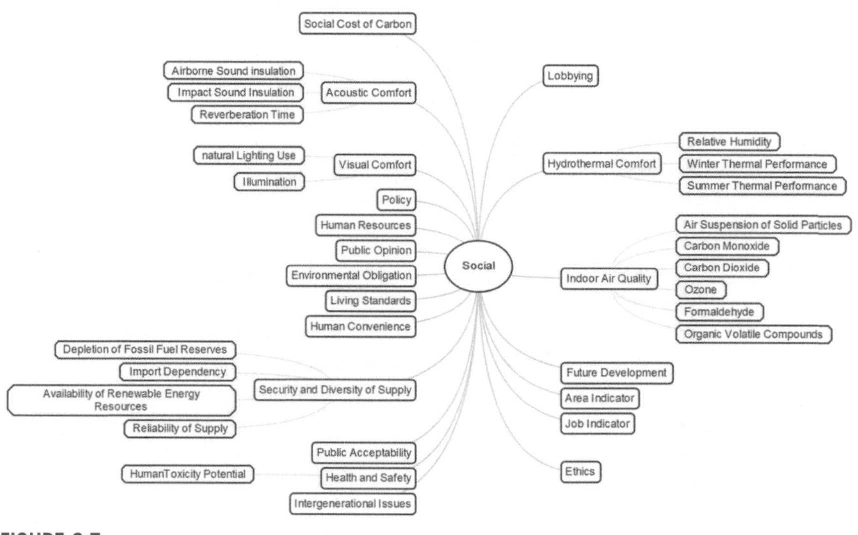

FIGURE 6.7

Concept map of the social category and the distribution of various parameters used in the literature to account for the social factor in sustainability assessment.

FIGURE 6.8

Social indicators proposed for the sustainability assessment in this model.

indicators to assess the global environmental impact potential, air pollution potential, and water pollution potential. LCA was also used to conduct the environmental aspect of this sustainability assessment. Gnanapragasam et al. (2010) had 10 environmental indicators to be part of his sustainability assessment model. He categorized them under ecological indicators and accounted for one-third of his sustainability indicators. He used 20 other indicators for social and technology-based categories.

Dincer and Acar (2016) considered global warming potential, social carbon cost, and acidification potential in their overview assessment on clean energy solutions for better sustainability. Furthermore, although they only used these three indicators, they also normalized rankings of potential nonair environmental impacts such as land use, water consumption, water quality of discharge, solid waste and ground contamination, and biodiversity. They also conducted a SWOT analysis for different energy options considered. Moreover, environmental category was taken into consideration in desalination supply chain performance assessment (Balfaqih et al., 2017) as well as sustainability assessment of groundwater remediation technologies (Da et al., 2017). Overall, the environmental category was vivid in most of the studies reviewed in this literature review. The exceptions were the works of Caliskan et al. (2011, 2012) as the focus of these studies were exergetic performance. LCA is the tool that was widely used in most of these studies to assess the environmental indicators. LCA is a tool used to investigate the environmental impacts of a product or a system while taking into account all the life cycle stages they go through. Fig. 6.3 illustrates the various environmental indicators used in the literature in various studies. Overall, the environmental category is well established and widely used in the assessment of sustainability. The tools to measure environmental impacts quantitatively through LCA and other tools make it readily accessible and convenient for the researcher to analyze environmental impacts of energy systems. Although Fig. 6.9 presents various environmental indicators, for this book, a comprehensive collection of environmental indicators of various types (climate change, air, water, etc.) is chosen. Thus, the proposed model is composed of 10 environmental indicators presented in Fig. 6.10. It is suggested that these 10 indicators are sufficient to provide an accurate environmental assessment of energy systems for sustainability purposes. Air pollution potential includes assessment of particulate matter, SO_2, CO, NO_2, O_3, and Pb.

Furthermore, water pollution potential includes eutrophication potential, freshwater ecotoxicity potential, and marine ecotoxicity potential. Ecological indicators refer to more general concepts such as availability, adaptability, environmental capacity, timeline, material rate, energy rate, ecological balance, and endurance. Global environmental impact potential represents the environmental impacts that effect the globe universally such as the global warming potential, stratospheric ozone depletion potential, and abiotic depletion potential. In addition, nonair impacts include land imprint, biodiversity, social carbon cost, solid waste, and quality of discharged water.

When analyzing the environmental impact that energy systems have, all types of impacts must be accounted to have an accurate environmental assessment. In this proposed model, air pollution is accounted by investigating the global warming potential, ozone depletion potential, air toxicity, and smog air resulting from the utilization of the system.

Water pollution is also accounted by measuring the water ecotoxicity, acidification potential, and eutrophication potential. Other impacts are also considered such as the land use, water consumption, and abiotic depletion.

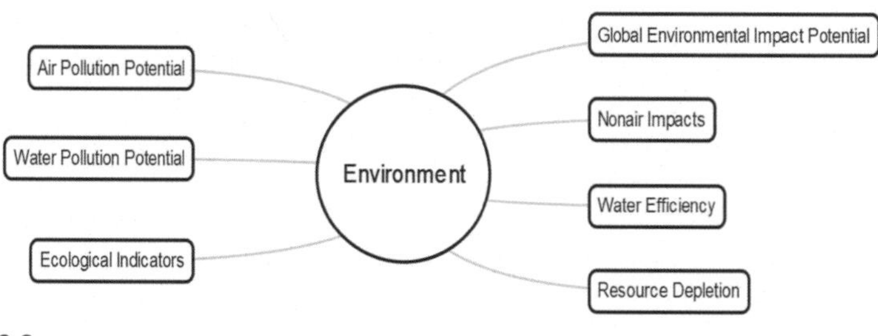

FIGURE 6.9

Concept map of the environment category and the distribution of various parameters used in the literature to account for the environmental factor in sustainability assessment.

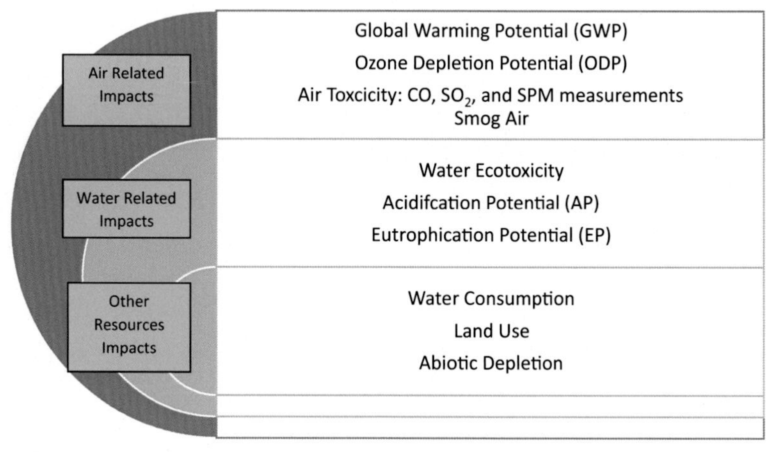

FIGURE 6.10

Environmental indicators adopted for the proposed sustainability assessment model highlighting air pollution, water pollution, and other forms of pollution as well.

6.2.7 Education

Although some studies investigated the social categories when assessing sustainability, the educational category has been majorly neglected. This category refers to the level of training and education available in a project or proposed energy system. Of course, the more educated and trained personnel available, the more sustainable the project and vice versa. So far, there has not been any study found, which dwelled into this category or even considered it as part of their sustainability assessment model. This book proposes investigating the educational aspect of the energy system by analyzing three main aspects illustrated in Fig. 6.11.

Further elaboration on the descriptions of these aspects will be presented in the following chapters. As this model provides a new integrated sustainability assessment model, internal parameters of a project or a system must also be considered when conducting the assessment. Thus, the educational category is an effective category to evaluate the general health and safety as well as the level of resource effectiveness and longevity of the project. Traditionally, training and innovation would be assessed under education. In this model, each concept is distinguished and assessed individually.

6.2.8 Size factor

The size factor is a category that mirrors the size of the energy system. The bigger the system, the more energy output can be yielded and vice versa. Bigger systems also need more maintenance and are more costly and thus it is a function of many factors to determine the sustainability of specific energy systems. Hacatoglu (2016) covered the same category in his assessment model taking into consideration the volume, mass, and area. This category is greatly neglected in the literature and is

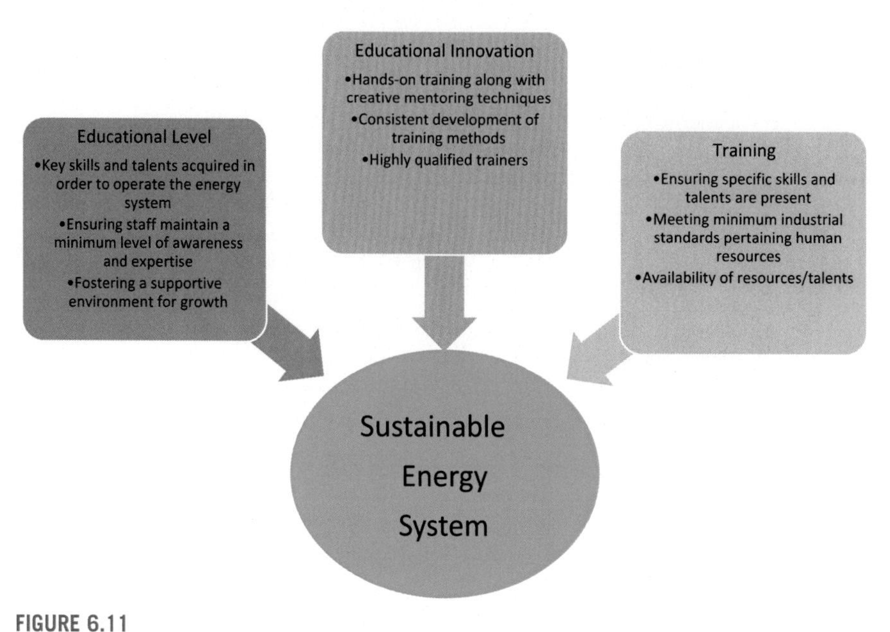

FIGURE 6.11

The relationship between education and sustainable development of energy systems.

not considered in most assessment models as such. This book proposes to take into account the size factor of energy systems by calculating the volume, mass, and area.

6.2.9 Summary

In summary, sustainability assessment models in the literature have focused on environmental, economic, and social aspects. However, to have a comprehensive and coherent assessment that is reliable and robust, other critical factors must be included as well. It is observed that studies in the literature tended to focus on certain categories while neglecting other categories. For example, some studies focused on the economic aspect of sustainability assessment, while others focused on environmental aspect. Furthermore, some studies have proposed novel assessment models that are integrated and some that are comprehensive. However, even these studies lack some major elements in certain categories. Furthermore, some studies are largely composed of qualitative assessment while the presence of quantitative assessment is limited, which decreases from its effectiveness and reliability.

6.3 Indicators

Indicators enable us to summarize, simplify, and condense complex and dynamic information to more meaningful and manageable information (Godfrey and Todd,

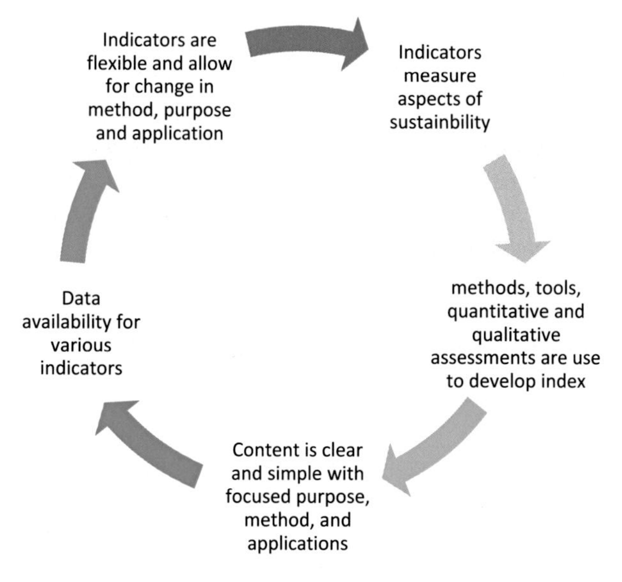

FIGURE 6.12

Indicator characterization and evaluation criteria based on Booysen (2002).

2001). Furthermore, some indicators might not be meaningful or useful if reference values such as thresholds are not provided (Lancker and Nijkamp, 2000). Some researchers used the "top-down" approach in their sustainability model by defining the target sustainability and consequently selecting indicators that would mirror the sustainability of the system. Others used the "bottom-up" approach, which requires systematic participation of stakeholders to ensure the development of indicators and the shaping of the sustainability framework simultaneously. Booysen (2002) introduced an assessment, which can be used to investigate the viability of indicators. He suggests that the classification and evaluation of indicators must be done based on the following general categories of measurement illustrated in Fig. 6.12. Furthermore, data need to be of high quality and accuracy. The techniques used to normalize, weight, or aggregate these values can also effect the results on sustainability assessment. Thus, methods and techniques need to be justified and chosen carefully.

6.4 Model development and framework

In this chapter, the approach of evaluation is described thoroughly. Sustainability is a complex and multidisciplinary concept, which requires detailed analysis and sensitive study to comprehensively understand its scope. Sustainability relates to the environment, economy, society, and other important factors. This chapter will describe the assessment methodology and elaborate more on the aggregation and weighting of data.

6.4.1 Methodology

To begin, it is important to note that sustainability is often only roughly measured, thus giving more of an estimation along with some economic and social indications. Until today, sustainability assessment does not have a universal standard, which makes this a crucial opportunity for scientists. Sustainability assessment of energy systems must be accurately and reliably comparable and measurable (De Vries et al., 2012). It is also vital to understand that sustainability assessment is a complex process with various inputs that need to be methodically quantified to obtain reliable, robust, and accurate information. Furthermore, the assessment ought to be comprehensive, taking into account the actual factors that relate to sustainability. Therefore, it is evident that a systematic method for identifying and generating sustainable solutions, which is shared universally, is still to be found.

This book aims to build on the research already conducted by scientists in this field by addressing these concerns. First, the model proposed is a comprehensive sustainability assessment model, which considers various aspects when assessing energy systems applications. The comprehensiveness of this model is novel and unprecedented. Fig. 6.13 illustrates the various categories and indicators used to assess sustainability. Second, the composition of the indicators was designed to be midway between the various research aspects, so that no aspect is neglected. Furthermore, each indicator has a purpose and measures a specific area of the energy system. All combined, a sustainability score is derived, which is accurate and meaningful for decision-making. It is important to mention that the proposed model was developed by identifying key and important parameters associated with each domain. Thus, double counting or under counting is minimized by selecting accurate and sufficient indicators per domain to accurately assess the sustainability of energy systems.

6.4.1.1 Energy aspect

Energy is a vital category of sustainability assessment as it reflects the systems' ability to produce reliable and useful energy that can be used for electric generation, heating or other applications. Various types of energy systems share this category and they vary among themselves in the efficiency and the production rate. For example, although renewable energy sources are mainly intermittent and rely on external factors for production, conventional energy sources are more reliable and would have higher production rates. Furthermore, efficiencies of energy systems vary greatly. Moreover, as the essence of energy systems is to provide useful energy for the growing demand of the world, it is only logical to consider the aspects around the energy production as crucial factors toward the sustainability assessment of these energy systems. In fact, energy-related impacts on sustainability are triggered by a rapidly increasing energy and global population demand (Dincer and Rosen, 2011).

The energy category is assessed using two indicators: efficiency and productivity. The score for this category is calculated as follows:

$$Y_{ER} = \left(\eta \times W_{\eta}\right) + \left(Y_{Pr} \times W_{Pr}\right) \tag{6.1}$$

where Y_{ER} refers to the total score of this category that is calculated by the addition of the scores of the two indicators. η refers to the score of the efficiency of the energy system; W_{η} refers to the weight that is given for this indicator. Y_{Pr} represents the score of the productivity of the energy system and W_{Pr} represents the weight associated with that indicator. When assessing the energy impact, the proposed model is confined to the following limitations and assumptions:

- This model is confined to energy systems that are in operation.
- The target energy efficiency is always assumed to be greater than the actual energy efficiency.
- When using linear aggregation, double counting may occur.

6.4.1.1.1 Energy efficiency

Efficiency refers to the level of performance that describes a process, in which the lowest amount of inputs is used to derive the greatest amount of outputs. It is a measurable concept, which is calculated by determining the ratio between the useful

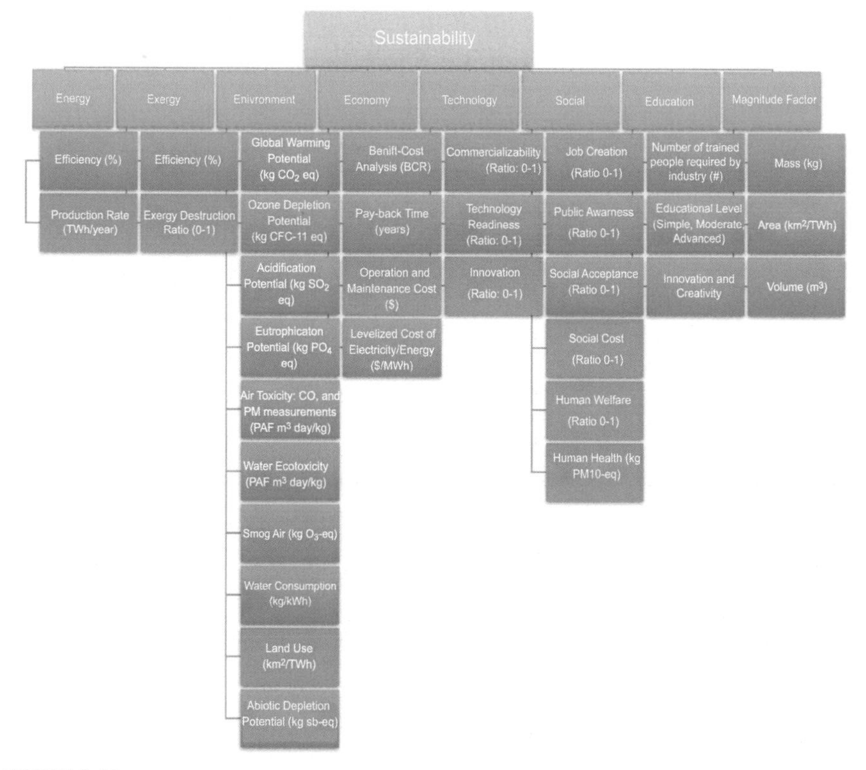

FIGURE 6.13

Outline of the proposed integrated sustainability assessment model. Domains are colored in green while indicators are blue in color.

output and the total input. In achieving the desired output, the concept of efficiency minimizes the waste of resource such as energy in any given energy systems. Indeed, efficient energy systems are better able to operate and produce useful energy in a sustainable manner. More efficient energy systems are also more environmentally benign and perform better economically while less efficient energy systems cause environmental pollution and are less economically favorable. Efficient energy systems have a direct impact on social trends in society such as maintaining a higher standard of living, including living in homes with running water and electricity as well as being mobile. Therefore, energy efficiency is suitable, important, and a reflective indicator to be used in sustainability assessment. Energy efficiency is directly correlated to the first law of thermodynamics. Thus, energy efficiency refers to the ratio of useful energy output in relation to the initial energy input. The actual efficiency of energy systems is always smaller than the upper thermodynamic efficiency limit. This is because the upper limit reflects the reversible reactions while all energy transformations include irreversibilities that decrease the efficiency below the targeted upper limit. As the target efficiency is always larger than the actual efficiency, the score for this energy efficiency indicator is calculated as follows (Hacatoglu, 2014):

$$\eta = \frac{1 - X_{ef(T)}}{1 - X_{ef}} \tag{6.2}$$

where $X_{ef(T)}$ refers to the target energy efficiency, which is the upper and reversible energy efficiency of the system. X_{ef} refers to the actual energy efficiency achieved by the system, including all the irreversibilities. The term $(1 - X_{ef(T)})$ refers to the minimum amount of unavailable energy while $(1 - X_{ef})$ refers to the actual unutilized incoming energy.

6.4.1.1.2 Production rate

Production rate compares the design value of the system. Energy systems that produce electricity at higher rates and with larger size are more favorable than the systems that have intermittent or low production rates. The score for this indicator is calculated as follows:

$$Y_{Pr} = \frac{X_{Pr}}{X_{Pr(T)}} \tag{6.3}$$

where X_{Pr} is the actual production rate of the energy system per hour. $X_{Pr(T)}$ is the upper target value for production rate in a year. It is calculated using the following equation:

$$X_{Pr(T)} = PR \times \dot{P} \tag{6.4}$$

where PR is the production rate (tonnes/hour) and $\dot{P}$ is the number of maximum operational hours in a year (hour/year). This number varies depending on the type of the system. For example, solar energy is intermittent and dependent on irradiance and sun availability while nuclear energy is independent of external weather factors. This way, each system is evaluated based on its internal value and function.

6.4.1.2 Exergy aspect

Exergy relates to the second law of thermodynamics, which is instrumental in providing meaningful and clearly comprehensible information toward environmental impacts. The most appropriate link between the environmental impact and the second law of thermodynamics has been namely exergy, mainly because exergy is a measurement of the departure of the state of a system from that of the environment (Kanoglu et al., 2009). The states of both the system and the environment both effect the degree of exergy. In practice, before exergy analysis, thermodynamic analysis of the system is conducted by the evaluation of mass and energy balances. Only energy conversion and transfers of the system are taken into consideration in the energy analysis while exergy analysis focuses on the quality of energy by measuring the degradation of energy or material in the system. Therefore, exergy analysis is associated to the first and second laws of thermodynamics and has the ability to identify the energy quality issues in the system or the work potential. Thus, exergy directly correlates with sustainability, as the assessment should also focus on the loss of energy quality along with the loss of energy itself in the system. Simply, exergy is an effective tool to measure the usefulness of an energy system and the degree of environmental impact an energy system has on the environment. Moreover, in order for energy systems to be considered smart, they need to be exergetically sound (Dincer and Acar, 2017). This implies that the system reduces exergy destruction to the minimum while simultaneously increases exergy efficiency to a maximum. The exergy category therefore, is assessed using two main indicators: efficiency and exergy destruction. The score of this category is calculated as follows:

$$Y_{EX} = (\psi \times W_{\psi}) + (Y_{ED} \times W_{ED}) \tag{6.5}$$

where Y_{EX} represents the total score for the exergy category. The score is calculated by adding both indicators. ψ represents the exergy efficiency of the system and W_{ψ} represents the allocated geometric weight for this indicator. Y_{ED} is the score of the exergy destruction indicator and W_{ED} is the geometric weight associated with it. When assessing the exergy impact, the proposed model is confined to the following limitations and assumptions:

- This model is confined to energy systems that are in operation.
- The target exergy efficiency is assumed to be greater than the actual exergy efficiency.
- When using linear aggregation, double counting may occur.
- Exergy-based indicators provide solutions for objective and robust measurements.
- The use of exergy in this sustainability assessment is limited to material exchanges, excluding exchanges with society.

6.4.1.2.1 Exergy efficiency

Exergy efficiency could be a more important indicator than energy efficiency as it usually gives a finer understanding of performance (Caliskan et al., 2011). Exergy efficiency highlights that losses and internal irreversibilities are to be assessed to improve performance. Higher exergy efficiency reflects higher energy quality used in the system, which consequently makes the system more sustainable while lower exergy efficiencies reflect energy losses and internal irreversible reactions, and thus low energy quality and worse sustainable score. Furthermore, exergy analysis enables the identification of energy degradation in an energy system and provides an accurate measure of the useful work that can be utilized from the system. Therefore, the exergy efficiency indicator is a useful tool for maximizing the benefit and efficiently using the resources.

Similar to the energy efficiency, the exergy efficiency's score is calculated as follows (Hacatoglu, 2014):

$$\psi = \frac{1 - X_{\psi \mathrm{ef(T)}}}{1 - X_{\psi \mathrm{ef}}} \tag{6.6}$$

where $X_{\psi \mathrm{ef(T)}}$ represents the reversible exergy efficiency of the system while $X_{\psi \mathrm{ef}}$ represents the actual exergy efficiency of the system.

6.4.1.2.2 Exergy destruction ratio

Exergy destruction is a measure of resource degradation. Although exergy efficiency measures the quality of exergy the system is harnessing, exergy destruction ratio is assessing the degraded resources and specifies the elements in the system where destruction is occurring. The exergy destruction ratio is calculated as follows:

$$\dot{E}x_{\mathrm{d}} = (1 - \psi)\dot{E}x_{\mathrm{in}} \tag{6.7}$$

where $\dot{E}x_{\mathrm{in}}$is the total exergy input to the system. For example, solar irradiance is the exergy input to solar energy applications while chemical and physical exergy of fossil fuels is the exergy input to fuel-based energy applications.

6.4.1.3 Environmental impact

Humans have been cherishing the concept of sustainability since the early civilization developments. Sustainable development however was environmentally friendly. The key milestone that created the gap between energy and the environment is the use of coal for energy production. The industrial revolution and the use of coal have transformed energy production forever because of the environmental impact it had through the massive emissions of greenhouse gases. The regular pollution caused by the coal revolution and later on followed by the oil revolution has rapidly triggered global warming and climate change. Fossil fuels and conventional energy sources have revolutionized the human lifestyle and social trends. Coal and oil (also known as black gold) have had a tremendous impact on the modern human civilization. However, the ease of lifestyle and comfort in standard of living came at the cost of environmental vulnerability of the planet. Environmental impacts could be local and specific to certain regions or global and widespread without geopolitical considerations. Furthermore, environmental impacts could also be short or long term. This category has been the most commonly used category in all sustainability assessment models. Energy systems are assessed according to their level of pollution and environmental impact. Various indicators are used to comprehensively reflect the impact of various energy systems on the environment. The score of this category is calculated as follows:

$$\begin{aligned} Y_{ENV} = {} & (Y_{GWP} \times W_{GWP}) + (Y_{ODP} \times W_{ODP}) + (Y_{AP} \times W_{AP}) + (Y_{EP} \times W_{EP}) \\ & + (Y_{AT} \times W_{AT}) + (Y_{WE} \times W_{WE}) + (Y_{SA} \times W_{SA}) + (Y_{WC} \times W_{WC}) \\ & + (Y_{LU} \times W_{LU}) + (Y_{ADP} \times W_{ADP}) \end{aligned} \tag{6.8}$$

where Y terms refer to the score for the indicators used while W terms refer to the weights assigned for the indicator. GWP refers to the global warming potential, ODP to the stratospheric ozone depletion potential, AP to the acidification potential, EP to the eutrophication potential, AT to air toxicity, WE to water ecotoxicity, SA to smog air, WC to water consumption, LU to the land use, and ADP to the abiotic depletion potential. These 10 indicators are carefully selected to account for all of the emissions and environmental impression that energy systems leave throughout manufacturing and operation of these systems. Further explanation follows for each indicator. When assessing the environmental footprint, the proposed model is confined to the following limitations and assumptions:

- The target environmental indicators have lower values than the actual indicators.
- Double counting may occur, as environmental indicators are not independent.
- This assessment methodology does not take into account all environmental indicators.

6.4.1.3.1 Global warming potential

Greenhouse gases contribute to the global climate change and global warming as they warm the earth by absorbing the incoming solar energy from the sun and trapping it within the atmosphere. Acting like a blanket insulating earth, they slow the rate at which energy escapes. Most common greenhouse gases that account for this include carbon dioxide (CO_2), methane (CH_4), and chlorofluorocarbons (CFCs). The element carbon is the common factor among the different greenhouse gases. Global warming potential (GWP) is a measure that was developed to compare the impact of different gases on the atmosphere. Specifically, it is a measure of how much energy is absorbed when 1 tonne of a specified gas is released to the atmosphere over a period, relative to the emission of 1 tonne of carbon dioxide. In this case, the larger the GWP, the more negative it is for the environment. CO_2 equivalence (CO_2-eq) is used as a measure for GWP. The time usually used for GWP is 100 years. Thus, the GWP indicator in this book considers the 100 year warming potential of all greenhouse gases throughout their life cycle. The following equation illustrates the calculation of the GWP score (Hacatoglu, 2014):

$$Y_{GWP} = \frac{X_{GWP(T)}}{X_{GWP}} \tag{6.9}$$

where X_{GWP} represents the actual greenhouse gas emissions for the period of 100 years. $X_{GWP(T)}$ represents the target value for this time period, which is the minimum greenhouse gas emissions, achieved by solely relying on renewable energy sources. This means, conventional energy sources such as fossil fuels are not considered in any stage of the energy production of the system. These values can be extracted by SimaPro as part of the LCA.

6.4.1.3.2 Stratospheric ozone depletion potential

Although life on the earth is impossible without light from the sun, solar radiations contain harmful ultraviolet (UV) rays. The ozone layer, located in the lower level of the earth's stratosphere, fortunately blocks these UV rays from reaching the earth's surface. Although some UV rays are beneficial, prolonged exposure is detrimental. Manmade CFCs have adversely affected the ozone layer. These CFCs react with the UV rays in the ozone layer and form chlorine (Cl) through a chain reaction. Chlorine then reacts with the ozone (O_3) and breaks its formation into (O_2). The breaking of the ozone layer causes a thinner ozone layer and a more opportunity for UV rays to infiltrate and reach the earth's surface. First used as working fluids in refrigerators, CFCs have been banned by the Montreal Protocol. However, CFCs have long residence time (46–1700 years) and old equipment that are still in use keep emitting these substances, which result in a very slow recovery for the ozone layer. CFC-11 is used to describe all ozone depleting substance emissions. The following equation illustrates the calculation of the ODP score (Hacatoglu, 2014):

$$Y_{ODP} = \frac{X_{ODP(T)}}{X_{ODP}} \tag{6.10}$$

where X_{ODP} represents the actual annual CFC-11 emissions per capita. $X_{ODP(T)}$ represents the limit of the CFC-11 emissions per capita. Setting this limit for the CFC-11 emissions per capita is a challenging task. This is because it acts as the target value and had it been set to zero, and then the solution would not be practical or realistic. To counteract this challenge, Hacatoglu (2016) proposed another way to calculate an acceptable amount of ozone depletion over the timescale of considering sustainability. The following is the proposed method of calculation (Goedkoop and Spriensma, 2000):

$$X_{ODP(T)} = \frac{O_3}{k_{Cl-O_3} \times f_{CFC-11} \times n_{Cl} \times POP_{world} \times t_{Sust}} \times \alpha_{ODP} \tag{6.11}$$

where k_{Cl-O_3} represents the relationship between the concentration of stratospheric chlorine and ozone depletion. f_{CFC-11} represents the fate factor for CFC-11 when emitted from the earth's surface. n_{Cl} represents the number of chlorine atoms in a single CFC-11 molecule. t_{Sust} is the timescale considered for the sustainability assessment. Although the timescale for sustainability assessment can range from 5 years to infinity, using an infinite value will yield in a zero target value. This reflects that there is no tolerance for stratospheric ozone depletion. The timescale used for this book is 100 years. This goes in line with the typical GWP calculation. SimaPro is used to conduct all LCAs to estimate the life cycle emissions and the impact of pollutants. Input data used to assess the ozone depletion is presented in Table 6.1.

6.4.1.3.3 Acidification potential

Acidification potential refers to the compounds that are precursors to acid rain. These include sulfur dioxide (SO_2), nitrogen oxides (NOx), nitrogen monoxide (NO), nitrogen dioxide (N_2O), and other various substances. Acidification potential is usually characterized by SO_2-equivalence. These acid gases are usually released into the atmosphere because of fuel combustion. On the other hand, newly constructed coal-fired power plants have a desulfurization technique to limit the SO_2 emissions to the environment. Acidification occurs with substances varying in their acid formation potential. The following equation illustrates the calculation of the AP score (Hacatoglu, 2014):

$$Y_{AP} = \frac{X_{AP(T)}}{X_{AP}} \tag{6.12}$$

where X_{AP} represents the calculated acidification potential (concentration of SO_2) in the local environment. $X_{AP(T)}$ is the latest set standard by EPA for the ambient air

Table 6.1 input parameters used in the life cycle assessment of the ozone depletion indicator for energy systems (Hacatoglu, 2014).

Parameter	Value
AreaS_ON	97281 km^2
ΔO_3	2%
$f_{CFC\text{-}11}$	2.8×10^{-9}
GHG	6.8 Gt CO_2 eq $year^{-1}$
$k_{Cl\text{-}O3}$	0.02
MATAI	\$69,300 $year^{-1}$
n_{Cl}	3
ODP	0.017
$Populations_{S\_ON}$	12.11 million
$Population_{WORLD}$	7 billion
R_{Sb}	4.63×1016 kg
t_{Sust}	100

quality, which is 190 μg m^{-3} (EPA, 2011). X_{AP} is calculated using the following equation:

$$X_{AP} = SO_{2,0} + \frac{SO_2}{Area_{Community} \times MH_{SO2}} \times \frac{\tau_{SO2}}{8760} \quad (6.13)$$

where $SO_{2,0}$, SO_2, τ_{SO2}, MH_{SO2} represent the background concentration, annualized life cycle emissions, residence time, and vertical mixing height of SO_2, respectively (Hacatoglu, 2014). For this book, $Area_{Community}$ represents the total area that a community of 160 households occupies.

6.4.1.3.4 Eutrophication potential

Eutrophication is a leading cause of impairment for many coastal marine and freshwater ecosystems. It is characterized by excessive growth of algae and plant due to increased availability of one or more limiting growth factors, which are needed to conduct photosynbook. Eutrophication is characterized by phosphate equivalence (PO_4-eq) in life cycle impact assessments. Eutrophication is often detrimental to plants and ecosystems and leads to the vulnerability of economic and social structures. The following equation illustrates the calculation of the EP score (Hacatoglu, 2014):

$$Y_{EP} = \frac{X_{EP(T)}}{X_{EP}} \quad (6.14)$$

where X_{EP} represents the actual life cycle emissions of PO_4 per capita per year. $X_{EP(T)}$ represents the target value, which is calculated using the following equation (Hacatoglu, 2014):

$$X_{EP(T)} = EP_{ref} \times \alpha_{EP} \quad (6.15)$$

where EP_{ref} represents the global annual per capita of PO_4 emissions and α_{EP} represents the adjustment factor.

6.4.1.3.5 Air toxicity

Air pollution is very common with the rise of industrial projects, innovative transportation means, and residential applications. A polluted air imposes a health and safety risk for inhabitants of this world. A number of substances will be assessed under this indicator. Fine particulate matter ($PM_{2.6}$) inflicts a health concern as they make their way to the lungs. Although the composition of particulate matter varies with regions, it generally indicates a mixture of solid particles and liquid droplets in the air. $PM_{2.6}$ refers to the particulate matter that is 2.6 μm in diameter or less. In Ontario, $PM_{2.6}$ is largely composed of nitrate and sulfate particles, elemental and organic carbon. Furthermore, although some $PM_{2.6}$ is carried into Ontario from the United States, it is primarily formed from chemical reactions, mainly from the transportation and residential applications. Another subindicator is the coarse particulate matter (PM_{10}), which is 10 μm or less. Carbon monoxide is also assessed in the toxicity of air. It results from incomplete combustion of fossil fuels. It is also a precursor for ground level ozone formation and smog air. The following equation illustrates the calculation of the AT score (Hacatoglu, 2014):

$$Y_{AE} = \frac{X_{AE(T)}}{X_{AE}} \tag{6.16}$$

where X_{AE} is the calculated air toxicity from the annual life cycle emissions. $X_{AE(T)}$ is the target emission value periodically published by EPA for various regions across the world. For the purpose of this study, the target value of the USA, which is 2.6 μg/ m^3 for $PM_{2.6}$ is used (Hacatoglu, 2014).

6.4.1.3.6 Water ecotoxicity

Similar to eutrophication, water ecotoxicity can cause harm to aquatic ecosystems. Emissions of toxic and lethal substances to water bodies are detrimental to the organisms and the sea life. The common unit to measure water ecotoxicity is measuring 1,4-dichlorobenzene (1,4-DCB). The following equation illustrates the calculation of the WE score (Hacatoglu, 2014):

$$Y_{WE} = \frac{X_{WE(T)}}{X_{WE}} \tag{6.17}$$

where X_{WE} represents the life cycle emissions of 1,4-DCB per capita per year. $X_{WE(T)}$ represents the target emissions to freshwater systems per capita per year. The upper target value is calculated as follows (Hacatoglu, 2014):

$$X_{WE(T)} = WE_{ref} \times \alpha_{WE} \tag{6.18}$$

where WE_{ref} represents the global annual per capita of 1,4-1DCB emissions to freshwater systems and α_{WE} is the adjustment factor.

6.4.1.3.7 Smog air

Smog air is mainly composed of ground level ozone and particulate matter formed near the troposphere. It usually appears as haze in the air due to the mixture of smoke, gases, and particles. Smog air has been linked to a number of adverse health and environmental impacts. Health impacts associated with smog air include thousands of premature deaths and increased hospital visits in several communities. Furthermore, adverse environmental impacts on vegetation, visibility, and structures have been traced to smog air. Warmer temperatures and hotter climate make a perfect ingredient for smog air and thus it is more common in the summer season. However, smog air is present in the winter as well. Smog's residence time in the troposphere is quiet short (1 h). The following equation illustrates the calculation of the SA score (Hacatoglu, 2014):

$$Y_{SA} = \frac{X_{SA(T)}}{X_{SA}} \tag{6.19}$$

where X_{SA} represents the calculated concentration of the ground level ozone (O_3). $X_{SA(T)}$ represents the upper threshold for ground level ozone set by the latest environmental protection agency standards, which is 160 $\mu g\ m^{-3}$ (EPA, 2011). The calculated concentration of the ground level ozone is calculated using the following equation:

$$X_{SA} = O_{3,0} + \frac{O_3}{\text{Area}_{\text{Community}} \times \text{MH}_{O3}} \times \frac{\tau_{O3}}{8760} \tag{6.20}$$

where $O_{3,0}$, O_3, τ_{O3}, and MH_{O3} represent the background concentration, annualized life cycle emissions, residence time, and vertical mixing height of O_3 respectively (Hacatoglu, 2014).

6.4.1.3.8 Water consumption

Water consumption is an important factor to consider when assessing sustainability of energy systems, especially in arid climates such as Australia, where water evaporation rates are quiet high. Although some LCAs have ignored the water requirements and availability for thermal systems, some have recently introduced them. Water consumption refers to the amount of water lost during the process of energy production. The following equation illustrates the calculation of the WC score:

$$Y_{WC} = \frac{X_{WC(T)}}{X_{WC}} \tag{6.21}$$

where X_{WC} represents the actual used water in the life cycle of the energy system and Table 6.2 shows different values that will be used for each system based on the works of Inhaber (2004). $X_{WC(T)}$ represents the target values for water consumption based on Spang et al. (2014).

Table 6.2 Water consumption of electricity generation from various sources (kg/kWh).

Energy source	X_{WC} (kg/kWh)
Photovoltaic	10
Hydro	36
Wind	1
Geothermal	12–300
Gas	78
Coal	78

Inhaber, H., 2004. Water use in renewable and conventional electricity production. Energy Sources 26(3), 309–322. https://doi.org/10.1080/00908310490266698.

6.4.1.3.9 Abiotic depletion potential

Abiotic depletion potential is a factor that is assessed in LCAs. It refers to the measure of the use of nonrenewable sources for energy production. The following equation illustrates the calculation of the ADP score (Hacatoglu, 2014):

$$Y_{\text{ADP}} = \frac{X_{\text{ADP(T)}}}{X_{\text{ADP}}} \tag{6.22}$$

where X_{ADP} represents the life cycle use of antimony and its equivalents per capita per year. $X_{\text{ADP(T)}}$ represents the annual sustainable antimony allocation. The threshold value is calculated using the following equation (Hacatoglu, 2014):

$$X_{\text{ADP(T)}} = \frac{R_{\text{Sb}}}{\text{POP}_{\text{world}} \times t_{\text{sust}}} \times \alpha_{\text{ADP}} \tag{6.23}$$

where R_{Sb} represents the recoverable reserves of antimony.

6.4.1.4 Economic impact

Economy is a critical category when assessing sustainability of energy systems. What does an economically sustainable energy system look like? This critical question must be addressed in any project before embarking on the execution journey. Furthermore, although conventional energy sources are relatively cheaper, renewable energy sources remain quiet expensive. However, improved economic planning and the progress toward cheaper renewable and clean energy are making the competition tougher between energy systems. Moreover, economic factors involved in the operation and design of energy conversion systems have brought the thermal energy storage for example to the forefront of its industry (Dincer and Rosen, 2007). Several thermal energy storage technologies are indeed present in the industry and are used side by side with on-site energy sources to economically buffer variable rates of supply and demand. In addition, an energy system is economically sustainable when they meet the following standards:

- The economic benefit of the energy generation outweighs operational, capital, and maintenance cost. Simply, the project is economically viable.
- Energy systems with shorter payback periods are preferred over systems with longer payback periods. This attracts investors.
- Lower levelized cost of energy/electricity. Energy available for everyone at a relatively lower cost.

In summary, energy systems are economically sustainable if they are profitable, serviced at lower cost for the consumer, and contain the elements of a successful business idea. The score of this category is calculated as follows:

$$Y_{\text{ECO}} = (Y_{\text{BCR}} \times W_{\text{BCR}}) + (Y_{\text{PBT}} \times W_{\text{PBT}}) + (Y_{\text{LCOE}} \times W_{\text{LCOE}}) \tag{6.24}$$

where Y_{BCR}, Y_{PBT}, and Y_{LCOE} refer to the scores of benefit-cost ratio, payback time, and the levelized cost of energy/electricity, respectively. "*W*" terms refer to the weight associated with each indicator. When assessing the economic impact, the proposed model is confined to the following limitations and assumptions:

- The benefit-cost ratio is confined to profitable projects only (i.e., no negative values).
- The payback time for energy projects is assumed to be between 0 and 23 years.
- LCOE target values are limited by values published by the US Energy Department.

6.4.1.4.1 Benefit-cost ratio

This indicator aims to explore the relationship between the benefit and cost of any proposed energy system. This indicator is informative both quantitatively and qualitatively as it analyzes all the possible benefits and costs. All benefits associated with an energy system are summed while all costs are subtracted. Although this analysis is routinely conducted in any business matter, it is novel to the sustainability assessment of energy systems. When conducting a cost-benefit analysis, results that are more accurate are achieved by analyzing the net present value (NPV) of all future costs and benefits. Simply, if NPV is negative, the project will never pay for itself and thus it is a financially losing project. However, if NPV is positive, the profits outweigh the costs and the project will pay for itself over time and eventually generate profits. The net present value is calculated using the following equation:

$$\mathrm{NPV} = \sum_{i=1}^{N} \left(\frac{\mathrm{PI}_i}{(1+r)^i} \right) - \mathrm{Cost}_0 \tag{6.25}$$

where PI_i represents the project's net income in a given year. N represents the number of years over which the project income occurs. r is the discount rate and Cost_0 is the project cost, typically assumed in the initial year (0). On the other hand, the benefit-cost ratio is another method of analyzing the benefits and costs of a given energy system. The following equations are used to determine the benefit-cost ratio:

$$\mathrm{BCR} = \frac{\ddot{P}}{\dot{N}} \tag{6.26}$$

where $\ddot{P}$ represents the present value of the net positive cash flow and $\dot{N}$ represents the present value of net negative cash flow.

6.4.1.4.2 Payback time

The payback period is an indicator used to assess the short- and long-term benefits of the proposed energy systems if any. Logically, energy systems with shorter payback periods are more economically favorable than those with longer payback periods. Thus, shorter payback period is associated with higher sustainability. The payback time refers to the time it takes in order for the project to recover all invested amounts and is usually expressed in years. Payback method does not take into account the time value of money different from the previous indicator (net present value or benefit-cost ratio). The calculation of the payback time is simple. The following equation is used to determine the payback period:

$$\mathrm{PBT} = \frac{\dddot{P}}{\mathrm{PCF}} \tag{6.27}$$

where $\dddot{P}$ represents the total project investment in ($) and PCF represents the periodic cash flow in ($/year). Table 6.3 shows the judgment criteria set to obtain the PBT score. Shorter payback time is advantageous and more attractive.

6.4.1.4.3 Levelized cost of electricity/energy

The levelized cost of electricity or energy (used interchangeably) refers to the cost of energy. It accounts for all lifetime costs of the system including operation, maintenance, construction, taxes, insurance, and other financial obligations of the project. They are then divided by the expected total energy outcome in the system's lifetime (kWh). Cost and benefit estimates are adjusted to account for inflation and are discounted to reflect the time value of the money. It is indeed a very valuable tool to compare different generation methods. Lower LCOE values resemble low energy cost, which in turn reflects back with high financial profit to the investors and vice versa.

$$\mathrm{LCOE} = \frac{\sum_{i=0}^{N}\left[\frac{I_i + O_i + F_i - TC_i}{(1+r)^i}\right]}{\sum_{i=0}^{N}\left[\frac{E_i}{(1+r)^i}\right]} \tag{6.28}$$

where I_i is investment costs in year I, O_i represent the operation and maintenance costs in year i, F_i represents the fuel costs in year i, TC_i represents the total tax credits in year i, E_i represents the energy generated in year i, r is the real discount rate, and N is the economic lifetime of the system.

Table 6.3 Scorecard for payback time.

Score	Payback time (PBT)
0.76–1	0 < PBT < 6
0.61–0.76	6 < PBT < 11
0.26–0.6	12 < PBT < 17
0–0.26	18 < PBT < 23

Table 6.4 The actual values for the LCOE of various ways of energy generations (Lazard, 2014).

Energy source	LCOE (US $/MWh)
Photovoltaic	80
Hydro	36
Geothermal	116
Gas	73
Coal	110
Nuclear	113

The value of the LCOE includes the capital cost average, fixed operation and maintenance cost average, variable operation and maintenance average, as well as the fuel cost average. The LCOE score is determined by the following equation:

$$Y_{\mathrm{LCOE}} = \frac{X_{\mathrm{LCOE(T)}}}{X_{\mathrm{LCOE}}} \tag{6.29}$$

where X_{LCOE} represents the actual LCOE of the energy system presented in Table 6.4. $X_{\mathrm{LCOE(T)}}$ represents the target value for the future and long-term LCOE for that system. For the purpose of this study, the values published by the US Energy Department for the LCOE for various energy systems in the year of 2040 will be used in the case studies for the values of $X_{\mathrm{LCOE(T)}}$.

6.4.1.4.4 Operation and maintenance cost

Energy systems that require frequent maintenance and operational follow-up are considered less sustainable as they are resource depleting, time consuming, and financially consuming. On the other hand, energy systems that function with minimal operational follow-up, or maintenance is more favorable and considered more sustainable. Operational and maintenance costs can be very high and thus for a system to reduce these costs, it is more sustainable. A value of 1 is assigned to systems that have low operational and maintenance cost and a value of 0 is assigned to systems that have high cost.

6.4.1.5 Technology

Technological indicators are used as part of this proposed assessment model and are considered important. Energy technology has transformed modern civilization, starting from the industrial revolution and the utilization of coal. Indeed, coal has revolutionized humans on the earth and introduced new applications in transportation, heating, and electricity generation. Furthermore, oil has also been a considerable milestone in human history as it introduced numerous novel technologies. Furthermore, the technological indicators assist in analyzing the performance, design, and production aspects of the energy system in question. Therefore, understanding the technological aspects of the proposed energy systems is important and vital to its sustainability. Commercializability and technology readiness are the two indicators that will be used to assess the technological category of this study. The score of this category is calculated as follows:

$$Y_{TECH} = (Y_{COMM} \times W_{COMM}) + (Y_{TR} \times W_{TR}) + (Y_{IN} \times W_{IN}) \quad (6.30)$$

where Y_{COMM}, Y_{TR} and Y_{IN} refer to the scores of commercializability, technology readiness, and innovation. "W" terms refer to the weight associated with each indicator. When assessing the technology domain, the proposed model is confined to the following limitations and assumptions:

- Technology is assumed to be regularly evolving and competitive.
- Assessment of this domain is open to all concepts regardless of location.

6.4.1.5.1 Commercializability

Although commercial viability is considered a weak point and sometime a threat of clean energy systems, there is an opportunity window for energy security and independence (Dincer and Acar, 2016). Commercialization refers to the potential for the energy system or technology to be commercially viable and enabling sustainable operation within the system. Mature and commercialized technologies are automatically considered more favorable than noncommercialized technologies that are still in the R&D commercially viable are considered less sustainable and a smaller value is appointed to them. Furthermore, multigenerational energy systems for example provide more commercial outputs, which increase their commercializability (Dincer and Acar, 2017). A value of 1 is assigned to systems that have reached the bankable assets phase and a value of 0 is assigned to systems that are not commercially viable yet (Hacatoglu, 2016 and Gnanapragasam et al., 2010). Fig. 6.14 shows the scale that assesses the commercializability of the energy system. A number of factors will be taken into consideration to determine the accurate level of the system including the

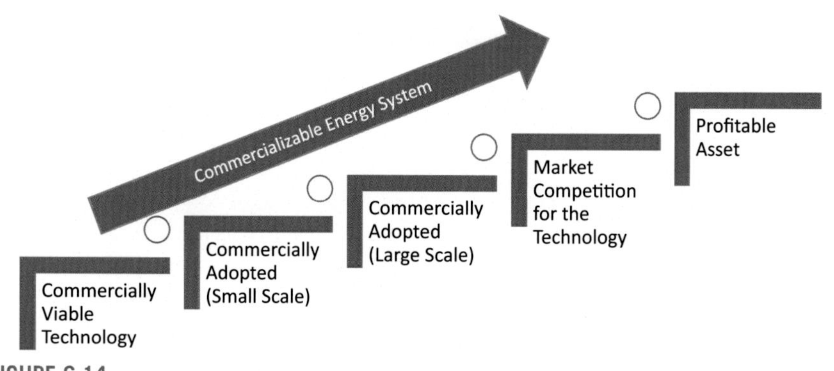

FIGURE 6.14

Scale of 0–1 to assess the commercializability of energy systems.

technical performance of the system, the stakeholder investment and acceptance of the technology, market opportunity, financial performance (cost and revenue), as well as the regulatory framework for that system. All in all, when a system is a bankable asset, it is considered most sustainable. If the system has some research progression and shows a commercially viability, it is the start toward more sustainable systems.

6.4.1.5.2 Technology readiness

Whether a technology is available or not in the current market is an important indicator to assess this category. Sustainable availability and readiness of the proposed energy system is important. Some technologies still need further research, experimentation, analysis, or legal work. On the other hand, some technologies have already been well established and are currently operational. A value of 1 is assigned to technologies or energy systems that are currently available in the market and commercially profitable and a value of 0 is assigned to technologies that do not exist in the market (Gnanapragasam et al., 2010). The value is typically higher for any technology that is available and ready. Fig. 6.15 illustrates the criteria that are used to assess the technology readiness level (Fig. 6.15).

6.4.1.5.3 Innovation

Innovation is an important criterion to be considered when analyzing technologies. Technologies that promote innovation and constantly enhance their development, research, and technology competitiveness are considered more favorable and sustainable. On the other hand, technologies that are stagnant and have limited enhancements to the technology are considered degraded and less sustainable. Innovation supported by scientific research as well commercialization yield in the birth of new technologies, which in turn flourish economic activities and lead to prosperous and enriched societies (Dincer and Acar, 2017). A value of 1 is given to the systems that incorporate innovative research and development, and a value of 0 is given to the systems that do not have innovative progression.

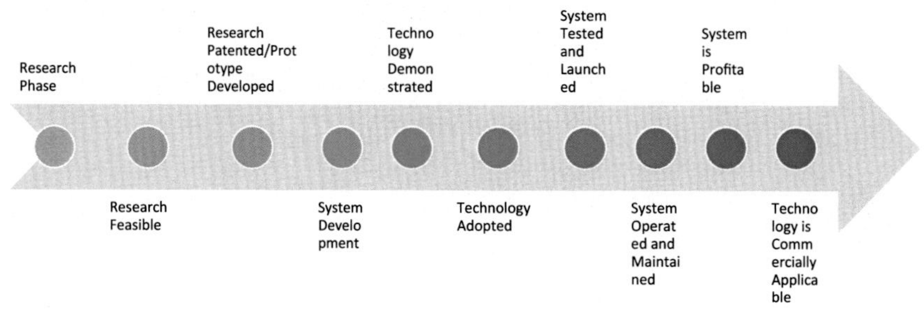

FIGURE 6.15

Scale from 0 to 1 to assess the readiness of the technology.

6.4.1.6 Social aspect

Social aspects of energy systems are very important for their sustainability. Social indicators help assess the impacts on the social system, which is composed of the beneficiaries of the energy system, whether directly or indirectly. In fact, proper utilization of renewable energy for example can have a direct impact socially and economically with further development of secure and sustainable energy supply (Dincer and Acar, 2016). On another important note, social morals and ethics is also a critical component of the social category as illustrated in Fig. 6.16. When addressing the concept of sustainability, adhering to a common set of principles and values can help govern the dynamics and the limits of energy systems. Therefore, ethical responsibility is essential to the social aspect of sustainability assessment as it provides a legal framework for sustainability. It is important to correctly identify and quantify the social indicators as they contribute to the acceptance and awareness socially. Fig. 6.16 illustrates some social indicators used to analyze this category and the interconnection between them. These elements are interconnected because job creation in a community causes awareness publicly and eventually leads to social acceptance. Furthermore, if a system is accepted socially, public awareness has the environment to flourish. On the other hand, if a system is rejected socially, the other two elements are adversely effected.

The score of this category is calculated as follows:

$$Y_{SOC} = (Y_{JC} \times W_{JC}) + (Y_{PA} \times W_{PA}) + (Y_{SA} \times W_{SA}) + (Y_{SC} \times W_{SC}) + (Y_{HW} \times W_{HW}) + (Y_{HH} \times W_{HH}) \tag{6.31}$$

where Y_{JC}, Y_{PA}, Y_{SA}, Y_{SC}, Y_{HW}, and Y_{HH} refer to the scores of job creation, public awareness, social acceptance, social cost, human welfare, and human health, respectively. "W" terms refer to the weight associated with each indicator. When assessing

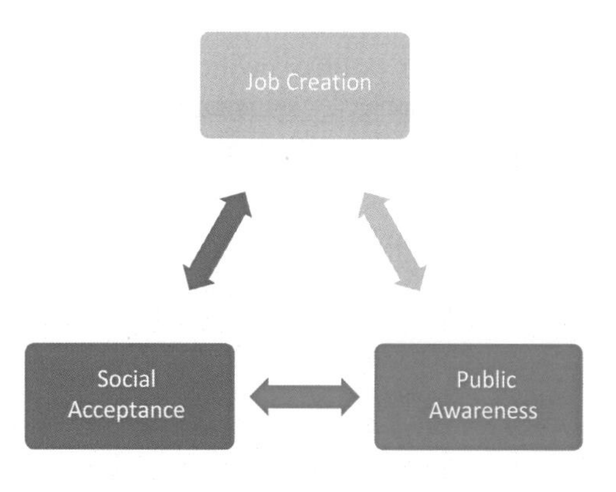

FIGURE 6.16

Illustration of the main indicators used for assessing the social category of the sustainability assessment model.

the social impression, the proposed model is confined to the following limitations and assumptions:

- The capacity for job creation is up to 69 jobs per megawatts as per IRENA and CEM (2014).
- Survey results are considered a reliable source of information. Therefore, the size and quality of surveys are assumed to be of high quality.
- LCA is used to assess the human health indicator.

6.4.1.6.1 Job creation

Energy systems have ever grown in the past few centuries, and they created many niches around them. When assessing an energy system, it is important to understand the social category behind the project and analyze the number of jobs that can be created to the local community or the larger region. Of course, more job creation is considered advantageous as that city prospers and attracts employees, talents from all over the surrounding regions. This increases the social life and the social activity in that local city, thus yielding in favorable results. It is considered sustainable when energy systems have high employment factor. The IRENA and CEM (2014) published a report on employment factors for wind and solar energy technologies. For the purpose of this study, the job creation factor is assessed based on the number of jobs created after each newly installed MW. The employment factor is presented in the units of (jobs/MW). Table 6.5 presents the judgment criteria for assessing this indicator.

6.4.1.6.2 Public awareness

Enhancing public knowledge and understanding about the issues that the energy industry is facing is vital to ensuring growth, energy sustainability, and security in our communities. Government programs, incentives, and other means of raising awareness all contribute toward creating an informed society. Indeed, innovative and coordinated awareness campaigns have had an impact on Scotland's perspective on renewable energy for example (McLaughlin and Smith, 2002). In this assessment model, it is considered that bigger positive public awareness is sustainable while smaller awareness is less sustainable. A value of 1 is assigned to systems that have big public awareness, and a value of 0 is assigned to systems that have little

Table 6.5 Judgment criteria for assessing the job creation indicator.

Score	Employment factor (Jobs/MW)
0–0.26	0 < employment factor < 9
0.26–0.6	10 < employment factor < 29
0.61–0.76	30 < employment factor < 49
0.76–1	60 < employment factor < 69

Source: Reproduced from IRENA and CEM, 2014.

awareness. Surveying is used to determine the public awareness of the project in interest.

6.4.1.6.3 Social acceptance

The power of the people is immense and thus for an energy system to be sustainable and operational, it must be accepted and perceived positively by society. For example, debates are still ongoing in several countries against wind energy, mainly because of its visual impacts on landscapes. Social acceptance therefore is an influential factor that could be a powerful barrier to the achievement of the energy targets of the system. This indicator has been neglected at the start of 1980s when policy programs were drafted at first. Later on, this factor surfaced to prove that it is essential before establishing an energy system in any locality. Therefore, successful energy systems are the ones that succeeded in integrating in the daily life of societies today (Dincer and Acar, 2017). Community acceptance goes hand in hand with social acceptance and is essential for sustainability assessment. For this model, the social acceptance score is determined by surveying the social acceptability of the project in interest. A value between 0 and 1 will be assigned to this indicator.

6.4.1.6.4 Social cost

This indicator brings a number of factors together in one value. The social cost is related to the economic category as well as the environmental, energy, and social categories. Energy systems usually come with a cost socially. This indicator has been assessed by calculating the social cost of carbon. This value helps to determine the monetary benefit/cost of regulations in reducing carbon emissions. Cost-free behavior of using fossil fuels has led to an addiction over these depleting resources.

6.4.1.6.5 Human welfare

Human welfare is a soft indicator that is used in this model to assess energy systems. Energy systems that take into consideration the welfare of society are more favorable and thus more sustainable. On the other hand, systems that are aversive to human welfare are considered less sustainable. A value of 1 is assigned to systems that have positive impact on human welfare, and a value of 0 is assigned to systems that have negative impacts.

6.4.1.6.6 Human health

With evolving technologies and innovative research, humans are exposed to various inputs that are constantly changing. The human health criteria are social indicators used to assess energy systems on the effects of any toxic substances on human health. Being exposed to various substances on a regular basis definitely has an impact. As a result, this indicator is considered important to comprehensively assess the sustainability of energy systems. A value of 1 is assigned to systems that have minimal human health impacts, and a value of 0 is assigned to systems that have high human health impacts.

6.4.1.7 Education

Education within the various stakeholders involved in the construction, operation, and maintenance of the energy system is vital to the sustainability and performance of that system. For example, staff that are more educated reflect more competent and skilled talents, which increase the sustainability score. On the other hand, poorly trained or educated staff could conduct the project in an unsustainable manner. Therefore, this category is calculated by assessing three main indicators. The score of this category is calculated as follows:

$$Y_{EDU} = (Y_{TRAIN} \times W_{TRAIN}) + (Y_{EL} \times W_{EL}) + (Y_{EI} \times W_{EI}) \tag{6.32}$$

where Y_{TRAIN}, Y_{EL}, and Y_{EI} refer to the score for the number of trained people required by the industry, educational level, and educational innovation. "*W*" terms refer to the weights associated with each indicator. When assessing the education domain, the proposed model is confined to the following limitations and assumptions:

- The educational level is confined to three categories as presented.
- The assessment of the subindicators could be qualitative and may be subjective.

6.4.1.7.1 Staff training

Industrial policies, increased health and safety standards, and general workplace awareness all contributed to creating healthier and more fruitful workplaces. The number of trained people required by the industry is an indicator that can help us to assess the educational category of energy systems. For example, if the energy system requires specific skilled staff, specific education, and rare talents, the system is perceived as less sustainable. On the other hand, when the systems' requirements on skills, trainings are widely available, then the system is more sustainable. A value of 1 is assigned to systems that meet the industrial standards of trainings and education, and a value of 0 is assigned to systems that do not abide by these standards.

6.4.1.7.2 Educational level of staff

The educational level is divided into three main categories: simple, moderate, and advanced. Advanced educational level is considered most sustainable while simple education is considered least sustainable. Table 6.6 shows the categorization of the different levels and the associated score.

Table 6.6 Educational level and respective score for each level.

Score	Educational level
0.76–1	Advanced
0.3–0.74	Moderate
0–0.29	Simple

6.4.1.7.3 Educational innovation

Inventing novel methods of learning, training, and educating is useful in this fast-growing society. Incorporating creativity, originality, and innovation in education is an indicator reflecting sustainable development and efficient planning. Energy systems that invest in innovation and creativity in their education stand out as most sustainable. A value of 1 is assigned to systems that effectively incorporate innovative and creative educational methods, and a value of 0 is assigned to systems that do not integrate such strategies in their educational plan if present.

6.4.1.8 Size factor

The size of the energy system is another important category to consider when considering their sustainability. Indeed, the categories of the proposed energy system could be a limiting factor. The size factor of the energy system in this sustainability assessment model will look at three main indicators: mass, land use, and volume. The score of this category is calculated as follows (Hacatoglu, 2014):

$$Y_{MF} = (Y_M \times W_M) + (Y_{LU} \times W_{LU}) + (Y_V \times W_V) \tag{6.33}$$

where Y_M, Y_{LU}, and Y_V refer to the score for mass, land use, and volume, respectively. "*W*" terms refer to the weights associated with each indicator. When assessing the size factor domain, the proposed model is confined to the following limitations and assumptions:

1. The size of the energy system can be a limiting factor depending on the application (i.e., mass is a limiting factor in mobile energy production systems).
2. The methodology assumes land use values as presented in Table 6.7.

6.4.1.8.1 Mass of energy system

The mass of the energy system is considered in this assessment by comparing the actual and target masses of the system. The following equation will be used to assess this indicator (Hacatoglu, 2014):

$$Y_M = \frac{X_{M(T)}}{X_M} \tag{6.34}$$

Table 6.7 Land use of various energy systems with no dual-purpose allocation (km^2/TWh).

Energy source	$X_{LU(T)}$ (km^2/TWh)	Source
Photovoltaic	28–64	Lackner and Sachs (2005)
Hydro	760	Evrendilek and Ertekin (2003)
Wind	72	(Gagnon et al., 2002)
Geothermal	18–74	(Bertani, 2005)

where X_M represents the actual mass of the system. $X_{M(T)}$ represents the target mass. Heavier systems are considered less sustainable. A value between 0 and 1 is assigned to mirror the appropriate condition of the system from this indicator's perspective.

6.4.1.8.2 Land use of energy system

Land use is another important indicator to assess energy systems sustainability. In specific, renewable energies are claimed to require large landmass, which interferes with agriculture and biodiversity. Photovoltaics and wind have similar land requirements. Moreover, while photovoltaics can be mounted on rooftops, thus providing a negligible footprint during use, wind turbines can be installed in agricultural lands. In both cases, the dual use of sites reduces the footprint caused by these technologies. The following equation illustrates the calculation of the LU score:

$$Y_{LU} = \frac{X_{LU(T)}}{X_{LU}} \tag{6.35}$$

where X_{LU} represents the actual land use of an energy system. $X_{LU(T)}$ represents the target land use from the literature presented in Table 6.7. Different references have been used to find the upper limit for each energy system.

6.4.1.8.3 Volume of energy system

The volume of the system is also taken into consideration to coherently assess the size factor of the system. Mobile energy production systems may be limited due to the volume of the energy system. Therefore, the following equation will be used to assess this indicator (Hacatoglu, 2014):

$$Y_V = \frac{X_{V(T)}}{X_V} \tag{6.36}$$

where X_V represents the actual volume of system. $X_{V(T)}$ represents the target volume. Bigger systems with larger volumetric values are considered less sustainable. A value between 0 and 1 is assigned to mirror the appropriate condition of the system from this indicator's perspective.

6.4.2 Multicriteria decision analysis

As the data collected based on this proposed model are already dimensionless, normalization is already accounted (Rowley et al., 2012). In multicriteria decision analysis, normalization refers to any process where diverse-unit cardinal scores are converted into a dimensionless numerical value with a common direction (Rowley et al., 2012). For this book, all variables are converted to values between 0 and 1 (score of 0 is less desirable than the score of 1). This step is usually a precursor to aggregation and weighting in various LCAs. Furthermore, in LCA, normalization is already embedded in the process during the life cycle impact assessment. For example, ReCiPe method adopts normalizations schemes based on the report of Sleeswijk (2007). In SimaPro, other methods are available and each one of them usually has a normalization method that is adopted as part of the LCA. Ratios and target values have been adopted to normalize the simulated and collected data and convert these values into dimensionless values. For example, indicator values ($X_{i,j}$) are normalized with respect to a target value ($X_{i,j,T}$) to generate a score for the subindicator ($Y_{i,j}$) as illustrated in the equations throughout this chapter. These target values are derived based on preferred upper/lower limits of thermodynamic, economic, and environmental thresholds. For example, consider Eq. (6.10) for the ozone depletion potential:

$$Y_{\mathrm{ODP}} = \frac{X_{\mathrm{ODP(T)}}}{X_{\mathrm{ODP}}} \tag{6.37}$$

If for instance, the target value for the ODP was 34.01 kg CFC-11 eq while the actual simulated value in SimaPro was 48.79 kg CFC-11 eq.; then the score for this indicator is calculated by dividing the target value by the actual value yielding in 0.696. This score is further processed by multiplying it with its weight depending on the scheme and later on by aggregating it with other values.

6.4.2.1 Compensability

Compensability refers to "the possibility of offsetting a disadvantage on some criteria by a sufficiently large advantage on another criterion" (Munda, 2005). For example, the energy efficiency is preferred to be as high as possible, but at the same time, the benefit-cost analysis is preferred to maintain its positive value. Increasing efficiency might require additional costs associated with technological upgrades and other miscellaneous costs. Therefore, in compensatory methods, a relatively lower efficiency is accepted while the benefit-cost analysis is of positive value. In noncompensatory methods, a higher efficiency is obtained regardless of the outcome in other criteria. When it comes to sustainability, the choice of algorithm requires that we define sustainability as weak or strong. Weak sustainability perspective enables the substitution of different forms of capital. In other words, the loss of rainforest, which is an ecological capital, may be offset by the financial gain capital gained from the development erected in its place. Strong sustainability perspective is the opposite, where certain natural capital is considered highly critical and cannot be substituted by manmade capital (Munda, 2005). Strong sustainability perspective is the preferred method mainly because it meets the accurate intent of the concept of sustainability. Therefore, for the purpose of this book, strong sustainability perspective is adopted, which entails using the noncompensatory method of aggregation. This is because each category is considered critical and important for assessing the sustainability of energy systems.

6.4.2.2 Data aggregation

As discussed earlier, noncompensatory aggregation is used in this model to ensure that each category is valued accordingly without undermining any important criteria. Once each indicator value is determined, they are all aggregated within one category to obtain a total value. For example, the economy category is assessed using four different indicators. These indicator values are grouped together to obtain the aggregated and total value for the economic category. Weighted arithmetic mean is used to aggregate values, where each indicator is assigned a specific weight, with all indicators totaling to 100%. Although determining weighting factors for each indicator might be controversial, many sustainability assessment models avoid the drawbacks around the subjectivity of the weighting by assuming equivalent weighting (Rowley et al., 2012). Therefore, equal weighting is used to aggregate values within one category. The following equation illustrates the weighted arithmetic mean calculation:

$$\mathrm{WAM}_{(Y,w)} = \sum_{i=1}^{n} w_i Y_i \tag{6.38}$$

where w_i is the weight associated with each indicator, Y_i represents the score of the indicator, and n represents the number of indicators in a given category. Linear aggregation assumes compensability among the indicators at this level (Juwana et al., 2012). This means that a very high value of an indicator can be compensated by a very low value of another indicator.

Another aggregation method used in this study is the weighted geometric mean. Once the scores are determined for both the indicators and subsequently the categories, these values need to be aggregated once more to come to a final aggregated score, representing the sustainability index of the energy system in study. Weighted geometric mean is a type of mean that indicates the central tendency of a group of values using the product of these values rather than their sum (arithmetic mean). With weighted geometric mean, some data points can contribute more to the final score than other data points in the model. The following equation illustrates the weighted geometric mean calculation:

$$\mathrm{WGM}_{(Y,w)} = \prod_{i=1}^{m} \left(Y_i^{w_i}\right) \tag{6.39}$$

where w_i is the weight associated with each category, Y_i represents the score of the category, and m represents the number of categories used in this model.

6.4.2.3 Weighting of data

Weighting is a very subjective tool, which may put the sustainability model at stake for biases and inaccuracies. One must acknowledge this weakness and try to minimize the subjectivity around weighting in various ways. In this book, five different characterization schemes are used to assess the sustainability of the case studies. These schemes include the individualist, hierarchist, egalitarian, panel, and equal weighting methods. Fig. 6.17 summarizes the different characterization methods and their differences. The hierarchist method stands out as moderate and balanced. Another common method to assign weighting factors is the panel method, where a panel of experts and stakeholders are consulted and weighting factors are distributed between the different categories in this model. Fig. 6.18 illustrates the process

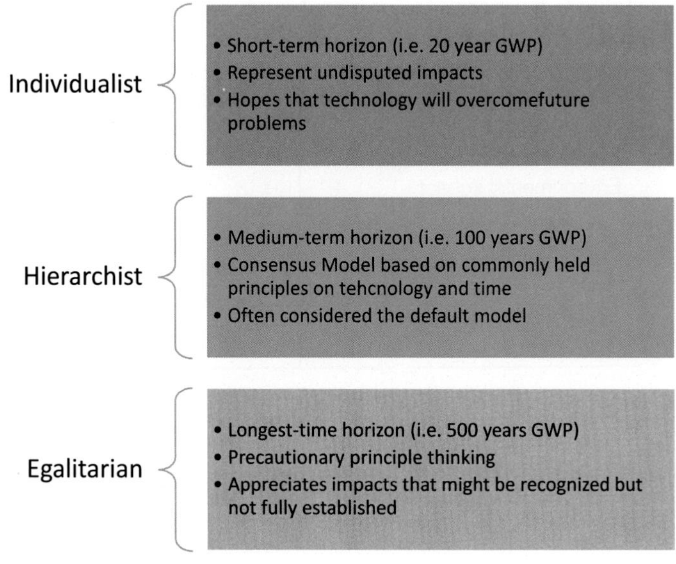

FIGURE 6.17

Summary of characterization methods used in weighting. Hierarchist method is used in this study for its balanced approach.

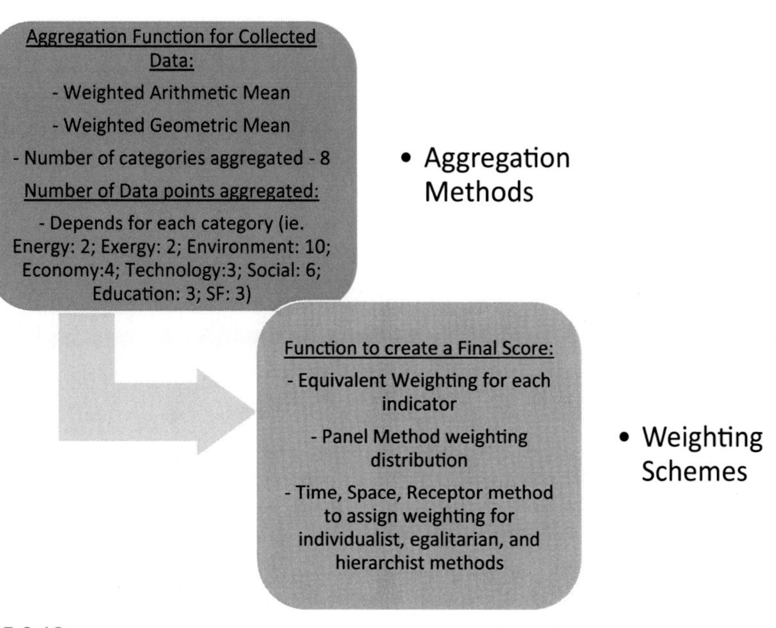

FIGURE 6.18

Data processing including aggregation and weighting of data. Indicators are aggregated and then weighting is applied to determine the final sustainability index.

Table 6.8 Priority factor distribution based on the panel method.

Category	Weight
Energy aspect	0.10
Exergy aspect	0.17
Environmental impact	0.18
Economic impact	0.12
Technology	0.16
Social aspect	0.14
Education	0.09
Size factor	0.06

the data goes through to come to a final sustainability index, which is a value between 0 and 1.

Moreover, Table 6.8 shows the different weights adopted for this model after conducting the panel review. One shortcoming of this approach is that the panel must have current and unbiased knowledge across enough of the impacts. As noncompensatory aggregation is used in this model, priority factors are needed to reflect the value for each category. Fig. 6.19 illustrates further details pertaining to the panel method.

Weighting will always be subjective in one way or another. It is important for the scientist to acknowledge this disadvantage and work toward minimizing the subjectivity as best possible. For the panel method, the panel composed academic and faculty experts in the field of sustainability from various Canadian universities including Ryerson University, University of Western Ontario, University of Toronto, and University of Ontario Institute of Technology.

Having a nonbiased composition is important in such assessments, which is the approach that this book tries to maintain throughout this research. Furthermore, all panel participants received the same information in the same format to reduce error and any associated bias. The procedure was simply direct rating and thus did not involve any complexities or lengthy discussions. Furthermore, in addition to the panel method, the individualist, egalitarian, hierarchist, and equal weighting schemes were conducted. The purpose behind that is to investigate the effect of weighting on the results. These schemes were also considered to reduce subjectivity of the results as much as possible.

To determine the relative priority factor of the indicators, a scale of 1—6 was adopted (1—very unimportant, 2—unimportant, 3—neutral, 4—important, 6—very important) with respect to time, space, and receptor criteria (Hacatoglu, 2014) (Fig. 6.19).

Table 6.9 shows the different schemes and the organization for evaluating the indicators to determine the priority factors. Each indicator used in this assessment was

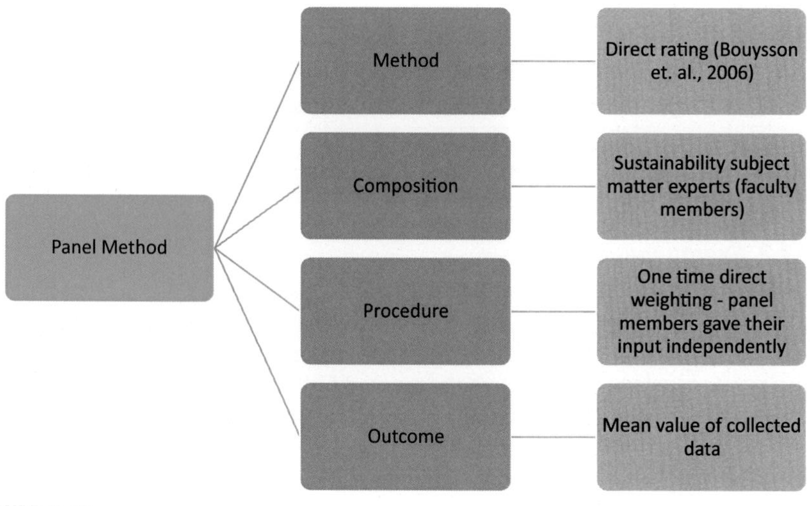

FIGURE 6.19

Using the panel method, various details are outlined pertaining how the panel was conducted and how the values were obtained.

Table 6.9 Difference between schemes with respect to time, space, and receptor.

Scheme	Time	Space	Receptor
Individualist	Short	Local	Humans
Egalitarian	Long	Global	Ecosystems
Hierarchist	Medium	Regional	Both

put to scale from 1 to 6 for time, space, and receptor. Weights are then determined by dividing the indicator's value by the sum of all values within the same category. For example, the energy efficiency's value after rating is conducted based on Table 6.9 is divided by the sum of the value of the energy efficiency in addition to the production rate.

6.4.3 Assessment indexes

For assessing the previously mentioned domains effectively, a number of indexes are introduced in this book to reflect the various parameters. Each domain explained in the methodology corresponds to an index, which reflects its influence over the sustainability index. The following is a brief summary of all assessment indexes introduced:

6.4.3.1 Energy index

Energy index refers to the energy aspect of the sustainability assessment. This includes various energetic parameters including the energy efficiency of the system, the energy storage rate, production rate, and other parameters obtained through the energy analysis, which is based on the first law of thermodynamics. Furthermore, this index accounts for the energy impact on other indexes as the environmental friendliness impact and the economic impact.

6.4.3.2 Exergy index

The exergy index reflects a number of exergetic parameters that are vital when assessing the sustainability of energy systems. Based on the second law of thermodynamics, exergy analysis is an integral tool to understand the quality of energy as well as point out where energy can be conserved further. Furthermore, the exergetic efficiency is taken into consideration by considering the total exergy of the desired output with respect to the total exergy input that is required. Exergy destruction ratio is also taken into account to assess this index comprehensively. In summary, the exergy aspect section of this chapter elaborates more about the importance of this index and provides details about its methodology.

6.4.3.3 Environmental friendliness index

The environmental friendliness index is a function of various environmental impacts of different categories including air-related, water-related, and other resources impacts. The degree of influence of the energy system on its surrounding environment is a critical factor when considering its sustainability. Furthermore, this index accounts for local impacts as well as global ones. A life cycle approach is adopted by following the CML2001 methodology. Moreover, this index illustrates the level of appropriateness of an energy system environmentally. In specific, the environmental impact section of this book dwells more on the methodology adopted in calculating this index. Preferred target values are selected by considering the lowest thresholds of various indicators.

6.4.3.4 Economic index

The economic index analyzes the financial aspect of energy systems by further investigating the benefit-cost ratio, payback time, maintenance, and operation costs as well as the levelized costs of electricity. Some of these parameters are local and of short-term nature while others are more global and of a long-term impact. Understanding the economic manifestations of energy systems is vital to understanding its sustainability. Therefore, this index covers a critical and important aspect of the assessment methodology. Furthermore, the energy sector is closely tied to economy, which makes this index valuable and an integral part to this assessment model.

6.4.3.5 Technology index

This index accounts for parameters associated with the technology itself. For example, the readiness of the technology is factored in this index. It also accounts for the commercializability of the technology along with its technological innovation level. The energy industry is continuously upgrading and therefore understanding the technological level at which the energy system has reached is important to assessing its sustainability. This index is further explained under the technology subheading of the methodology. Moreover, specific scales have been developed in this book to track and quantify the technological parameters.

6.4.3.6 Social index

The social index refers to the social aspects when assessing the sustainability of energy systems. It takes into consideration job creation, which is a local and a rewarding parameter. It also accounts for social awareness and social acceptance. As technologies evolve, societies respond differently to various technologies. In addition, some cultures might have certain preferences toward an energy system over another. Therefore, understanding the social value behind energy systems is important. Furthermore, the social index measures the human health, human welfare, and social costs behind energy systems. These parameters give value for this index as it gives clear and practical social results.

6.4.3.7 Educational index

The educational index reflects the level of education, training, best practices, and innovation in educational methods adopted in various systems. It is logical that staff who are more competent, well trained, and highly educated produce more efficient and effective products. Thus, this index accounts for these various educational parameters, which are mentioned in detail in the methodology. Moreover, certain systems require specific talents, which are rare to find or specific skills, which are limited. This limiting factor is also important to know when assessing the sustainability of energy systems.

6.4.3.8 Sizing index

This index really refers to capacity index in conjunction with the energy use (in kW) based on the volume, mass and land utilized. The volume and mass parameters are associated with mobile applications such as vehicles. The land use is associated with stationary applications such as a power plant. The sizing index reflects the magnitude of the energy system and consequently gives an accurate understanding when assessing its sustainability. This index is further explained in the methodology section of this book.

6.5 Closing remarks

In closing, the assessment of energy systems from a sustainability perspective requires a multidisciplinary paradigm shift that takes into account all aspects of the energy system besides its thermodynamic and environmental performance. This chapter introduced the sustainability model proposed for assessment of energy systems including all equations, assumptions, and references. Although this model presents a comprehensive and integrated model for sustainability assessment, it remains subjective to a certain degree and can be modified or enhanced further, especially the social domain. Moreover, although some aspects can be assessed quantitatively, some factors are confined to qualitative assessments, which need to be developed to accurately investigate energy systems. In addition, the economic, social, and environmental domains are critical indexes for this assessment in addition to the thermodynamic energy and exergy indexes. Furthermore, there are various methods for aggregating values and arranging the priority factors including the panel method, equal method, hierarchist, egalitarian, and individualist methods, which differ in time, space, and receptor focus.

CHAPTER

Case studies

7

7.1 Case study 1: sustainability assessment of Ontario's energy sector

In this study, Ontario's energy portfolio is assessed from a sustainability perspective. The assessment focuses on the primary domains of sustainability comprising the environmental, economic, and social aspects. Furthermore, the energy portfolio is explored via four main sectors, which are the industrial, residential, commercial, and transportation sectors. The industrial sector demands the highest amount of energy in Ontario followed by the transportation sector. Natural gas is a significant contributor to the industrial, residential, and commercial sectors, yielding substantial greenhouse gas (GHG) emissions and an adverse environmental impact. Furthermore, motor gasoline and diesel are the primary energy sources for the transportation sector, both of which are fossil-based fuels with undesirable environmental impact. In addition, the iron and steel, petroleum refining, and pulp and paper industries constitute the top three industries requiring energy in the industrial sector, consuming approximately 342 PJ, which is almost 70% of the total consumed energy in this sector. Furthermore, Enbridge supplies to approximately 70% of all Ontario customers while Union Gas supplies to approximately 40%. Nearly 70% of the total natural gas volume is supplied to nonresidential applications while the remaining 30% are supplied for residential use. In Ontario, electricity is distributed to consumers across Ontario through 71 local distribution companies. Hydro One Networks Inc. alone accounts for 70% of all electricity assets and 71% of all liabilities in Ontario. Additionally, residential customers are the main customers for Hydro One Networks in electric consumption and revenue. Nuclear and small hydro projects yield longer job years per MW during construction and installation phases, as they compete to provide jobs to Ontarians. On the other hand, natural gas is least favorable socially and does not have a competitive job creation capacity. Additionally, Canadians have clear support for renewable and clean energy sources while opposition is evident toward nuclear and fossil-based fuels.

Energy Sustainability. https://doi.org/10.1016/B978-0-12-819556-7.00007-3

7.1.1 Introduction

In 2017 Ontario's energy generation exceeded 3000 PJ out of which 30% was generated from natural gas, 11% from nuclear, 4.7% from hydro, and around 1% from other renewable energy sources (IESO, 2016). The current fuel mix is dominated by fossil-based fuels and nuclear energy contributing to nearly 90% of the total energy generation. Although coal generation has been phased out from Ontario, natural gas has become the top ranking energy source for GHG emissions. On the other hand, natural gas is a primary source of energy in Ontario, especially for industrial processes and for space heating in the residential and commercial sectors. This is due to its low cost and high availability.

In Ontario, the energy mix while diverse is seldom studied or analyzed. This study addresses this limitation by exploring the energy portfolio of Ontario as a whole and analyzing it from a sustainability perspective by assessing the environmental impact, economic performance, and the social aspect of this portfolio in the province of Ontario. Furthermore, this study highlights important elements of the energy sector and explores its sustainability.

When it comes to energy, sustainability assessment of electricity generation and renewable energy have dominated the literature (Evans et al., 2009). Most work seeks to quantify parameters such as emissions (Lenzen, 2008), costs (Kammen and Pacca, 2004), and energy payback periods (Kato et al., 1998). Thorough life cycle analyses (LCA) have been also conducted for various individual energy generation technologies. The role of hydrogen storage in renewable energy management for Ontario was studied by Ozbilen et al. (2012). For instance, Mallia and Lewis (2013) conducted a specific LCA for electricity-generating facilities in Ontario in 2008. GHG intensity for each fuel type is calculated along with the LCA GHG intensity of the Ontario grid as a whole. Furthermore, another life cycle cost and economic assessment of biochar-based bioenergy production in Northwestern Ontario was conducted by Homagain et al. (2016). In addition, the energy demand and the sustainability of different bioethanol production processes from sugar beet can be significantly improved by introduction of new technologies (Santek et al., 2010). Furthermore, Szekeres and Jeswiet (2018) conducted an interesting research evaluating the effects of technological development and electricity price reductions on adoption of residential heat pumps in Ontario. The use of heat pumps introduces more effective and increasingly efficient heating solutions to residential applications. Moreover, this solution allows for energy savings and GHG emission reductions. Besides, the focus on renewable energy in Ontario has also been another topic of interest in the literature. Furthermore, White et al. (2013) emphasized the policy consistency in the role of governments toward renewable energy solutions. Moreover, Dampier et al. (2016) investigated the potential local and regional induced economic impact of an energy policy change in rural Northwestern Ontario. Their study also studied socioeconomic footprints of various energy sources. On a separate front, Zhang et al. (2010) investigated the life cycle emissions and cost of producing electricity from coal, natural gas, and wood pellets in Ontario; whereas, Bloemhof (2017) assessed the consumer benefits in the Ontario residential retail natural gas market. In addition, Martire et al. (2015) investigated the enhancement of environmental sustainability in energy production by assessing the carrying capacity of forest resources. Moreover, Ontario's Green Energy and Green Economy Act have been the case study for Winfield and Dolter (2014) as they examined the energy, economic and environmental discourses as well as their policy impacts.

Although the literature has addressed certain aspects of sustainability individually and in a disintegrated fashion, this study analyzes Ontario's energy as a whole and assesses the environmental impact, economic performance and social aspect of this portfolio. Furthermore, this study examines the energy portfolio by dissecting each sector including the industrial, residential, and commercial; and transportation sectors. This dissection goes in line with published reports and plans of the Ontario Energy Board, Ministry of Energy and other energy regulatory stakeholders. Therefore, this study makes it more accessible for policy makers and researchers to understand the dynamics of energy in Ontario.

This case analyzes Ontario's energy from a sustainability perspective. The objective of this paper is to assess the current energy portfolio for the province of Ontario based on three main domain areas including, environment, economy and social aspects. Ontario's energy comes from a diverse, yet uneven base of energy sources including fossil-based fuels, renewable energy sources, and nuclear energy sources. In fact, transportation fuel and natural gas make up 77% of Ontario's energy use in 2017. Furthermore, electricity use makes up 20% of Ontario's energy use in that same year (Environmental Commissioner of Ontario, 2017). Therefore, the specific objectives of this case study are as follows:

- Analyze the current energy supply and demand in Ontario by quantifying the energy dynamics in each major sector: industrial, residential, commercial, and transportation sectors.
- Investigate the sustainability of energy sources in Ontario.
- Assess the environmental impact of Ontario's energy by gauging the annual GHG emissions in various aspects of each major sector.
- Estimate the economic performance of the energy portfolio by examining the financial aspects of natural gas and electricity generation in Ontario.
- Evaluate the social implications of the energy dynamics in Ontario by exploring social indicators such as job creation, perceived cost of electricity, social acceptance, and public awareness associated with the energy sector.

7.1.2 Sustainability assessment

Ontario's energy portfolio is very complex as primary sources of energy include a diverse variety of fuels. Fig. 7.1 illustrates the high-level summary of Ontario's energy sector. Consequently, this study explores the primary sources, and transformation of these sources up to the end uses of energy. Overall, the methodology and approach used in this study to assess Ontario's energy portfolio is unique and distinct, allowing for more effective and accurate assessment of the subject in study.

The major domains of sustainability are investigated in association with the end-use sector. This way, the primary sources, end uses, and the domains of sustainability are integrated in a cohesive and logical configuration. Furthermore, assessing Ontario's energy in this manner allows for more concrete sustainability results, analysis, and recommendations. Moreover, this configuration makes it easier for policy makers and researchers in the energy sector to fathom the sustainability aspects of Ontario's energy in a practical and scientific fashion. Fig. 7.2 demonstrates the relationship between sustainability of energy in Ontario and the three major domains used in this assessment.

Ontario's energy use remained constant between 2007 and 2017 despite an 8% population growth. This is due to energy efficient improvements, which also account for 17% economic growth. Therefore, per capita energy use decreased by 7% and

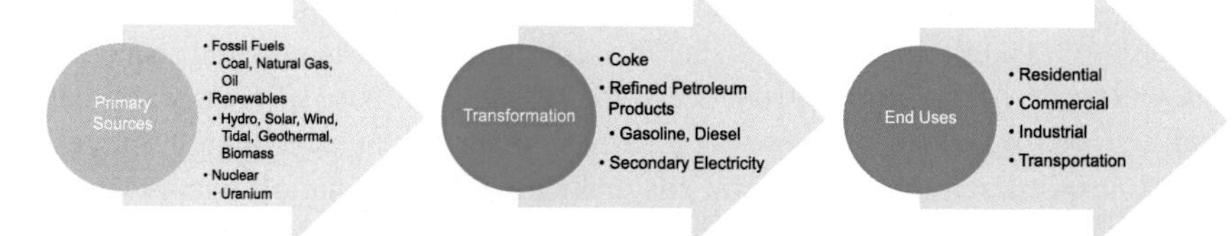

FIGURE 7.1

The flow of energy from primary sources to end uses in Ontario.

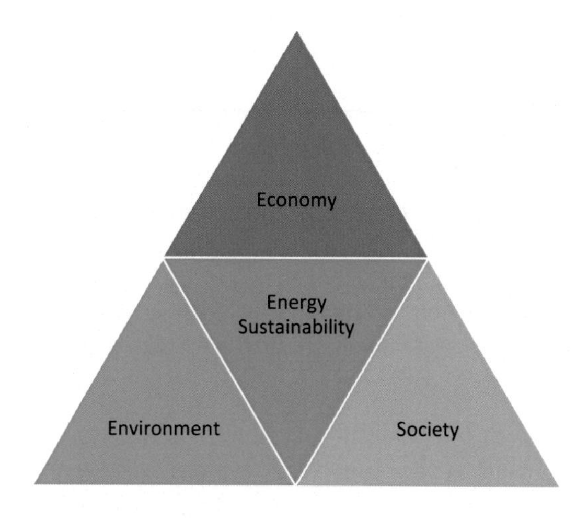

FIGURE 7.2

Sustainability in relationship to the three major domains assessed in this study.

energy use per dollar of GDP also decreased by 17% (Environmental Commissioner of Ontario, 2017).

Furthermore, Ontario's electricity primarily comes from nuclear operations, followed by hydropower. Natural gas accounts for only 10% of electricity generation followed by 9% from intermittent renewable energy including solar and wind (Environmental Commissioner of Ontario, 2017). Energy is supplied to five main sectors categorized as residential, commercial, industrial, transportation, and agriculture. Fig. 7.3 illustrates Ontario's energy demand for 2017 highlighting the residential, commercial, and industrial sectors and the sources of energy that supply each sector. The data used come from the National Energy Board using the reference case published in Canada's Energy Future Report (NEB, 2017).

It is clear that natural gas is the primary source of energy for these sectors accounting for approximately 42% of the total energy demand for these sectors. In fact, it is the main source for the residential energy demand producing 77% of the annual demand. Electricity, refined petroleum products (RPP), and liquefied petroleum gases (LPG) are the other main sources of energy meeting Ontario's demand. RPP includes aviation fuel, diesel, gasoline, heavy fuel oil, kerosene, light fuel, asphalt, naphtha, and other lubricants. Industrial energy demand is the largest of all sectors accounting for 1127 PJ.

This sector does not rely on renewable energy sources. Rather, it is mainly supplied by fossil-based fuels such as RPP, LPG, natural gas, coal, coke, and coke over gas as well as petroleum coke and still gas. Residential energy demand is primarily supplied by natural gas for the purpose of heating and cooling and electricity for power purposes. Similarly, commercial energy demand mainly comprises heating, cooling, and electricity. Thus, natural gas and electricity supply this demand almost

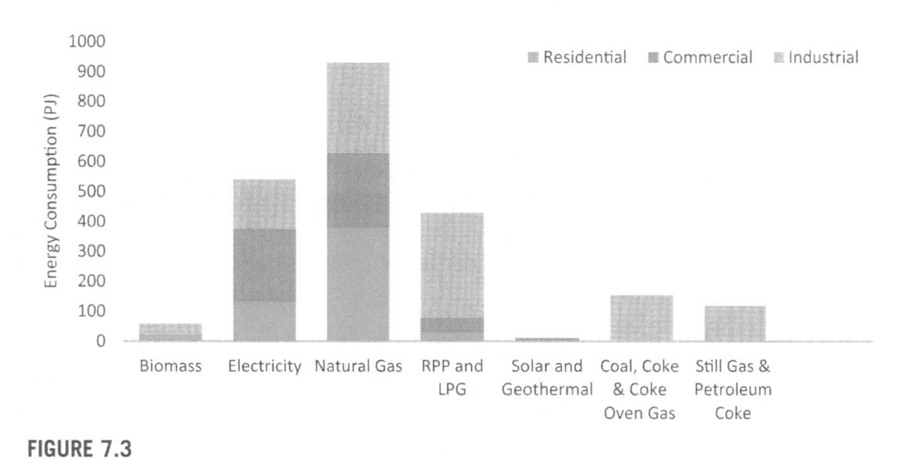

FIGURE 7.3

Ontario's energy demand for 2017 highlighting three sectors and the main energy sources.

Data from National Energy Board, 2017. Canada's Adoption of Renewable Power Sources. NEB, pp. 1–32. ISSN No. 2371-6804.

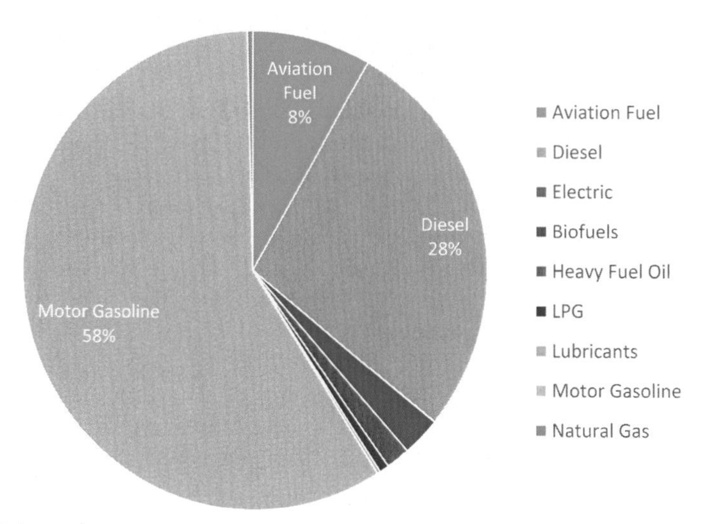

FIGURE 7.4

Ontario's energy demand for 2017 for the transportation sector and the main sources of energy supplying this demand.

Data from National Energy Board, 2017. Canada's Adoption of Renewable Power Sources. NEB, pp. 1–32. ISSN No. 2371-6804.

equally. Solar and geothermal are used insignificantly for buildings in the residential and commercial applications. Fig. 7.4 illustrates the transportation sector and the main sources of energy supplying Ontario's demand for 2017. The transportation sector is the second largest sector in energy use after the industrial sector.

The transportation sector accounts for approximately 30% of the energy demand for Ontario. As observed in Fig. 7.4, the sources supplying this third are in fact fossil-based with major environmental impacts. Motor gasoline supplies 713 PJ of energy to meet the transportation demand. Diesel is the second major source in this sector accounting for 28% of all supplied energy. Electric fuel supplies an insignificant amount of 0.2 PJ to this sector. Fig. 7.7 presents an overall summary of the main sectors and the composition of Ontario's energy demand in 2017.

From Fig. 7.5, it is clear that the industrial and transportation sectors make up more than 70% of Ontario's energy needs. Energy demands from the residential and commercial sectors are similar because they both are mainly concerned with supply energy to buildings and institutions.

7.1.3 Environmental impact assessment

This section will investigate the environmental impact for the industrial, transportation residential, and commercial sectors. GHG emissions will be the primary indicator used in this assessment, and the units of Mt of CO_2e will be the unit of measure.

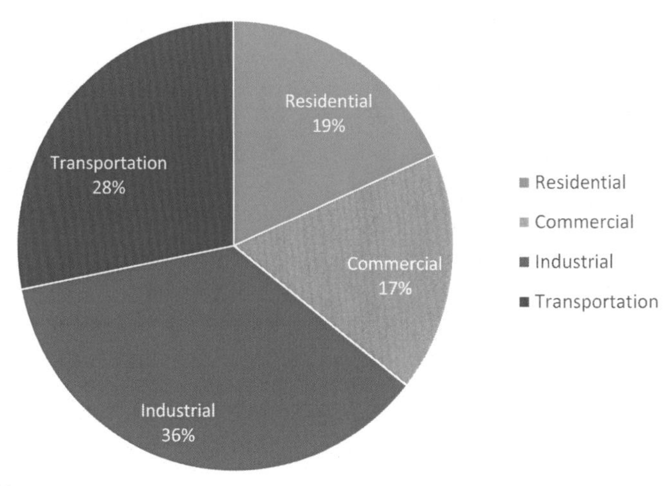

FIGURE 7.5

Ontario's energy demand for 2017 for the four main sectors.

Data from National Energy Board, 2017. Canada's Adoption of Renewable Power Sources. NEB, pp. 1–32. ISSN No. 2371-6804.

7.1.3.1 Industrial sector

The industrial sector comprises various manufacturing disciplines, highlighted by the iron and steel, pulp and paper, petroleum refining, and other manufacturing industries. These industries definitely have a significant footmark economically as it does environmentally and socially. As previously discussed, this sector demands the highest amount of energy on annual basis. Although higher energy demand reflects industrial growth, consequently introducing economic growth, energy conservation is necessary to reduce GHG emissions. Fig. 7.6 shows the recorded energy consumption of the major industries along with the respective total of GHG emissions.

Iron and steel, petroleum refineries and pulp and paper industries are the main industries that compose the highest energy consumption. However, the pulp and paper industries' GHG emission is much less compared to their consumption when compared to the iron and steel industries' GHG emissions. On the other hand, construction, cement factories, mining, and forestry are industries that consume the least amount of energy and thus their environmental footprint is quiet less.

The iron and steel, petroleum refining, and pulp and paper industries constitute the top three industries requiring energy in this sector, consuming approximately 342 PJ, which is almost 70% of the total consumed energy in this sector. A collection of other manufacturing industries also accumulates to represent 27% of the total consumed energy. However, the iron and steel industry has the highest record in GHG emissions with record of 11.3 Mt of CO_2e. Petroleum refining industry is the second major pollutant in this sector, accounting for 18% of total emissions. Furthermore, although the pulp and paper industry is the third top industry in energy

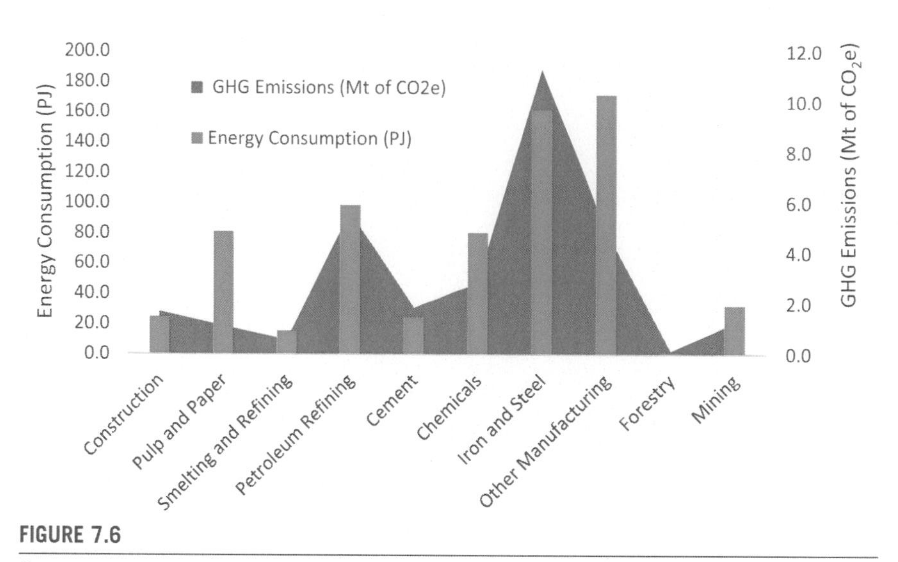

FIGURE 7.6

Energy consumption and GHG emissions in the industrial sector per industry for the year 2017.

Data from Natural Resources Canada, 2017.

use, its emissions are quiet low, emitting only 1.1 Mt of CO_2e. Similarly, other manufacturing industries make up 27% of the energy demand, yet their GHG emission share is only 17.7% of all total emissions in this sector. To understand these GHG emissions better, Fig. 7.7 illustrates the GHG emissions per each energy source for this sector.

The highest emissions come from the use of natural gas as this source is the top energy source used in this sector. Natural gas alone account for 12 Mt of CO_2e, which make up approximately 38% of all emission in this sector. Furthermore, while electricity is the second major source for the industrial sector, its emissions were excluded in this report. Moreover, still gas and petroleum coke also account for higher emissions, totaling almost 19% of all emissions. Certain sources have discontinued in Ontario's industrial sector including coal, coke and coke oven gas, wood waste, and pulping liquor.

Ontario's GDP grew by 2.7% in 2017. Service-producing industries contributed more to Ontario's GDP growth than goods-producing industries. Although retail trade advanced 3.7%, wholesale trade grew 3.9% in 2017. Increased computer systems design and related services resulted in increasing the professional, scientific, and technical services by 3.7%. Furthermore, manufacturing increased 1.8% as increases in output were reported by food products, plastic products, vehicle parts, pharmaceuticals, manufacturing, and medicines.

In summary, natural gas is the main source of energy for this sector, which accounts for the high GHG emissions. Energy conservation and the phase out of natural gas must be addressed with reliable and practical solutions. Furthermore, energy

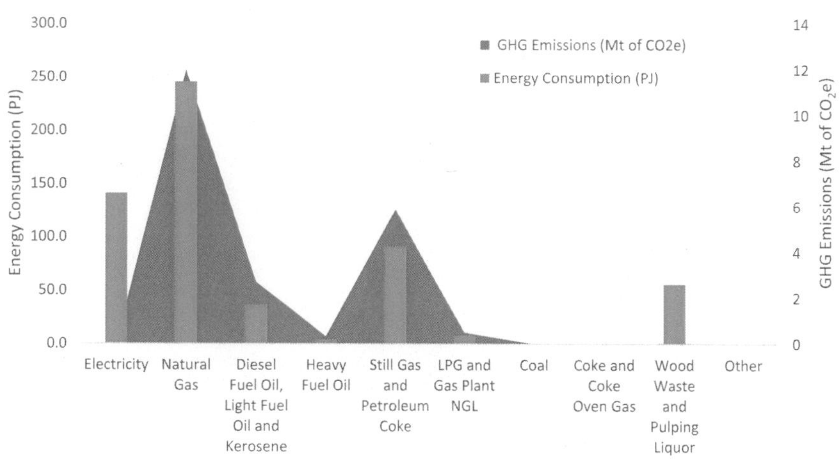

FIGURE 7.7

Energy consumption and GHG emissions in the industrial sector per energy source for the year 2017.

Data from Natural Resources Canada, 2017.

conservation programs specific to the iron and steel, petroleum refining, and pulp and paper industries are recommended to encourage less GHG emissions and the use of environmentally benign and friendlier energy options.

7.1.3.2 Transportation sector

The transportation sector is the second highest sector requiring energy in Ontario. Transportation is directly connected with lifestyle and has become an essential part of modern societies. This sector represents personal travel, as well as movement of trucks carrying goods and providing service. It also includes all types of public transportation in Ontario. Fig. 7.8 illustrates the energy consumption and GHG emission in this sector for the year 2017.

It is evident that passenger light trucks and heavy trucks constitute most of the energy demand in this sector, accounting for approximately 70% of the total energy demand. Consequently, their respective GHG emissions are also significant totaling 27 Mt of CO_2e. This is because this sector is highly fueled by fossil-based fuels as illustrated in Fig. 7.2. Diesel and motor gasoline are the primary sources of energy supplied to this sector. In addition, this sector is multidisciplined and can be associated to a wide variety of fields. Fig. 7.9 demonstrates various modes of transportation used in Ontario and their associated percentage of GHG emissions for the year 2017. Cars are responsible for 27% of the total emissions in this sector. On the other hand, passenger light trucks and heavy trucks make up 41% of the total emissions.

7.1.3.3 Residential sector

The residential sector in Ontario requires energy mainly for heating during the winter, cooling during the summer, and electricity throughout for power and utility

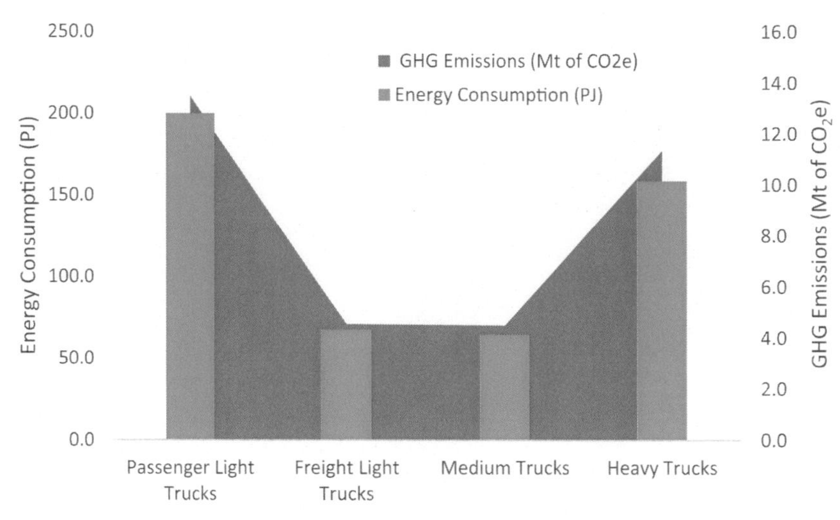

FIGURE 7.8

Energy consumption and GHG emissions in the transportation sector per truck type for the year 2017.

Data from Statistics Canada, 2017. Report on Energy Supply and Demand in Canada, 1990–2016, CANSIM (Table 128-0016), Ottawa.

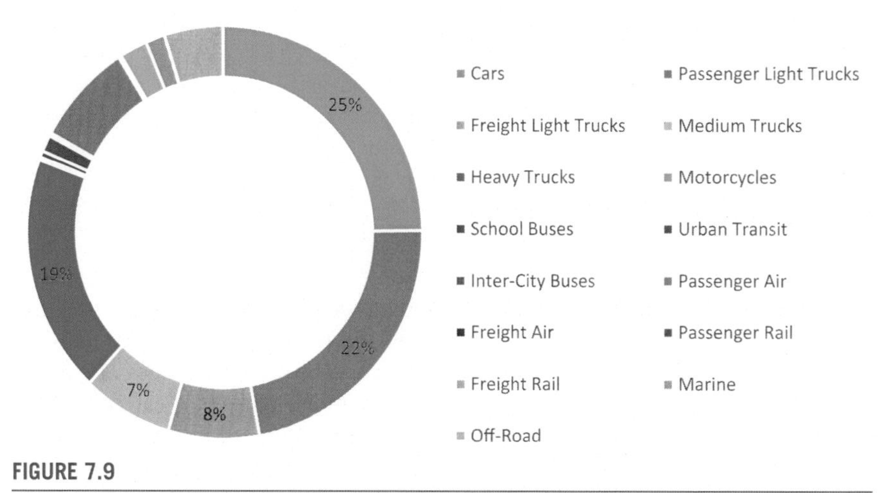

FIGURE 7.9

GHG emissions in the transportation sector by transportation mode for the year 2017.

Data from Statistics Canada, 2017. Report on Energy Supply and Demand in Canada, 1990–2016, CANSIM (Table 128-0016), Ottawa.

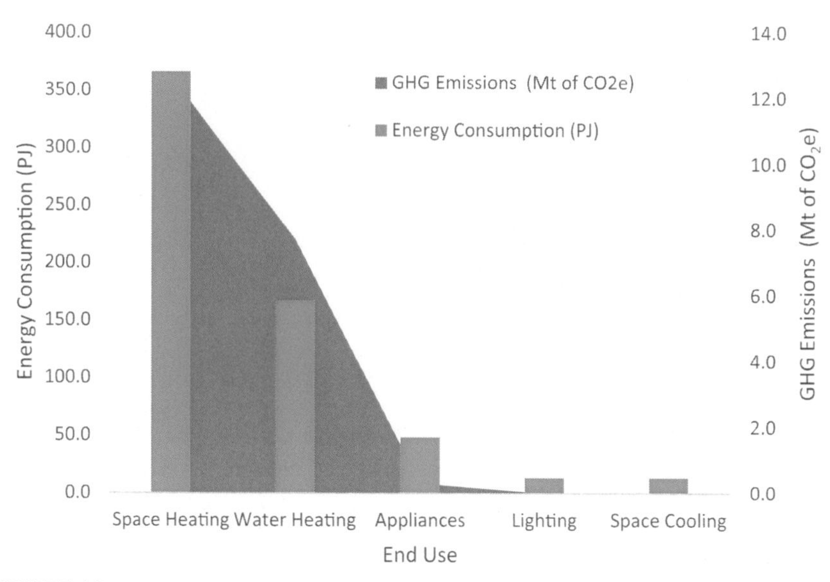

FIGURE 7.10

Energy consumption and GHG emissions in the residential sector by end use for the year 2017.

Data from Environment and Climate Change Canada, 2017. National Inventory Report 1990–2016: Greenhouse Gas Sources and Sinks in Canada, Ottawa.

purposes. This sector represents a wide variety of different types of accommodations as well as a wide variety of uses within the building. Fig. 7.10 presents the major end uses in the residential sector in Ontario for the year 2017 along with the respective energy consumption and GHG emissions for each end use.

As expected, space heating is the primary commodity in this sector requiring the most energy. The energy demand for this commodity is 70% of the total energy demand in this sector. Furthermore, the GHG emissions from this commodity are extremely high, accounting for 71% of all total GHG emissions from this sector. On the other hand, space cooling requires a very insignificant amount of energy, yielding zero GHG emissions. Similarly, lighting and appliances does not contribute negatively to the environment. However, water heating is the second major commodity in this sector, requiring 178 PJ and emitting 7.7 Mt of CO_2e in 2017. As discussed before, the main reason behind the high emissions resulting from the space heating is because of the use of natural gas. As shown in Fig. 7.11, natural gas is the primary source of energy supplied to this sector, followed by electricity.

Other sources include propane. GHG emissions clearly peak at the natural gas source because of its negative environmental impacts. Natural gas alone accounts for 88% of all emissions in this sector. In other words, decreasing this source or replacing it by other environmentally benign sources to provide space heating to residential buildings will optimize the environmental performance of this sector.

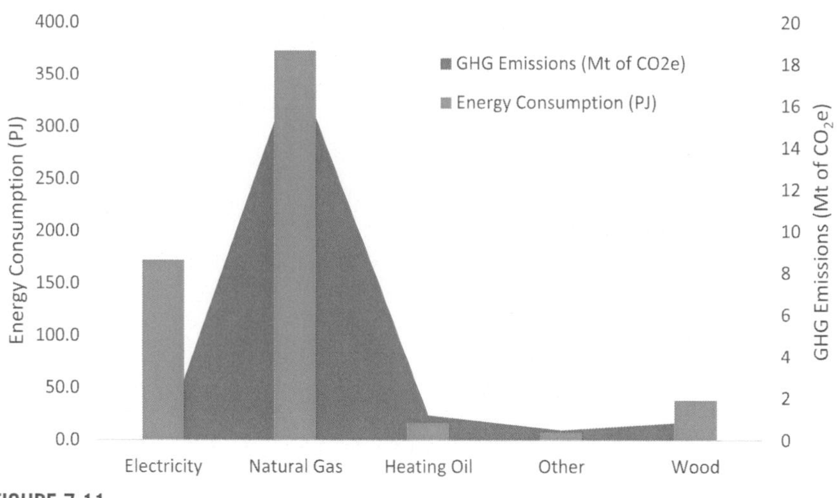

FIGURE 7.11

Energy consumption and GHG emissions in the residential sector by energy source for the year 2017.

Data from Environment and Climate Change Canada, 2017. National Inventory Report 1990–2016: Greenhouse Gas Sources and Sinks in Canada, Ottawa.

Heating oil, wood, and propane are also used for residential applications. In this figure, the GHG emissions from electricity are excluded from the published report. Furthermore, the GHG emissions of propane and wood are 1.4 Mt of CO_2e while the emissions for the heating oil are only 1.2 Mt of CO_2e.

Moreover, the residential sector comprises many types of buildings. As the space heating was the primary commodity in this sector to demand energy, it is investigated further. Fig. 7.12 shows the space heating energy use by building type along with the respective GHG emissions for each type. These values exclude the electricity consumption in the process of space heating. Single detached buildings consume the most energy amount while mobile homes and single attached homes consume significantly less energy. Furthermore, apartments are conservative consumers of energy as well. The reason behind the significant energy consumption for single detached homes is primarily their quantity.

Single detached buildings are the primary building types in this sector requiring energy for space heating. This may be because there are a huge number of these buildings or that they may be inefficient. Furthermore, both approaches may apply as well. Making up almost 77% of all energy demand for this commodity in Ontario, single detached buildings are responsible for most GHG emissions resulting from this commodity. Alternatively, apartments demand only 13.3% of the total energy demand for this commodity. Similarly, this could reflect the higher efficiencies of these types of buildings or that there are fewer number of them in Ontario. Moreover, both scenarios may be effective to a certain extent in this case. On the other hand,

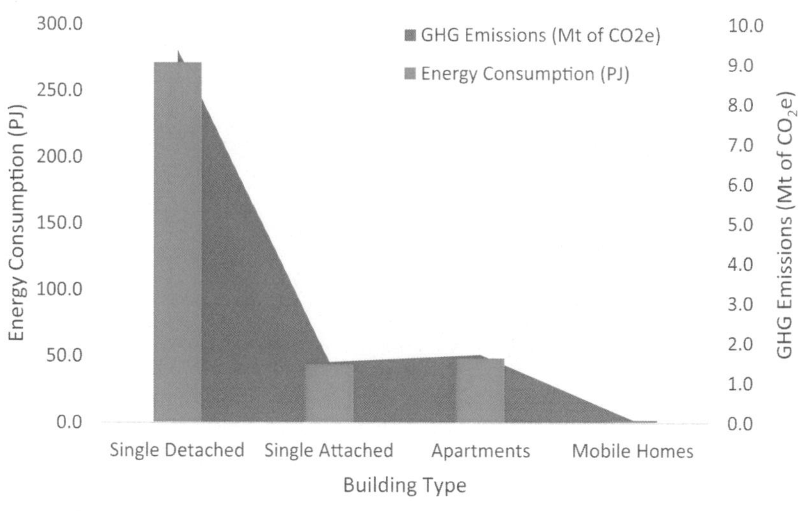

FIGURE 7.12

Energy consumption and GHG emissions in the space-heating commodity by building type for the year 2017.

Data from Natural Resources Canada, 2017.

water heating was the second highest commodity requiring energy in this sector in Ontario. To understand it better, Fig. 7.13 demonstrates the energy consumption and GHG emissions associated with each building type for this commodity.

The single detached building type remains the primary consumer among all other building types for this commodity. The GHG emission associated with this type of building is also the highest. A different trend exists between single attached buildings and apartments compared to their trend in Fig. 7.11. In this case, apartments require more energy than single attached buildings to meet the demand for water heating.

7.1.3.4 Commercial sector

The commercial sector represents a wide variety of institutions and commercial buildings. Therefore, it is very similar to the residential sector in commodities and energy needs. Fig. 7.14 shows the main commodities in this sector and their respective energy use and GHG emissions in Ontario for the year 2017. Similar to the residential sector, space heating is the primary commodity in this sector that requires energy as it represents more than 70% of the total energy demand. Consequently, the GHG emission associated with this commodity is significantly high at 10.3 Mt of CO_2e. Auxiliary motors, lighting, street lighting space cooling, and auxiliary equipment have very insignificant GHG emissions. Water heating however has a slight environmental footprint in this sector. Similar to previous reports, the GHG emission in this figure does not account for electricity generation.

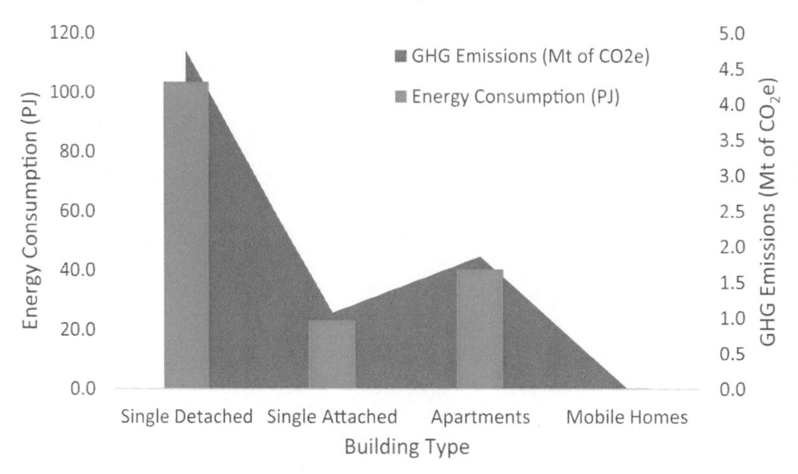

FIGURE 7.13

Energy consumption and GHG emissions in the water-heating commodity by building type for the year 2017.

Data from Environment and Climate Change Canada, 2017. National Inventory Report 1990–2016: Greenhouse Gas Sources and Sinks in Canada, Ottawa.

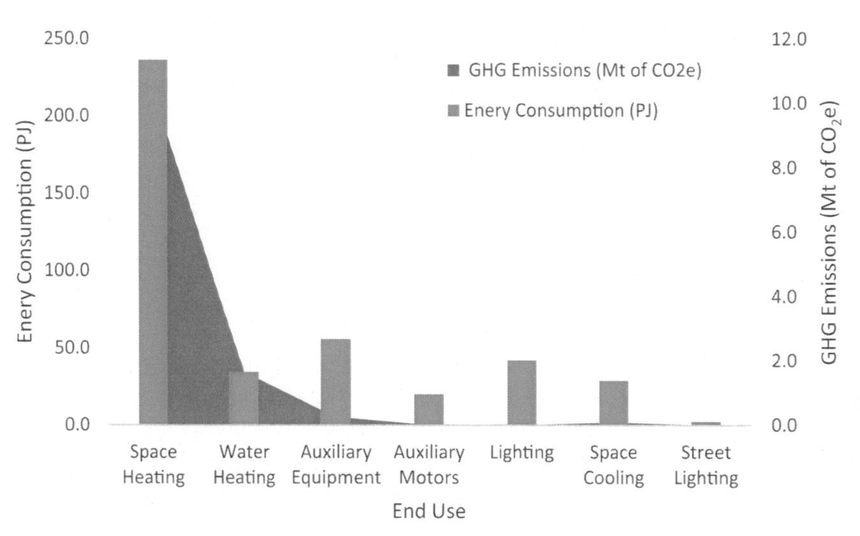

FIGURE 7.14

Energy consumption and GHG emissions in the commercial sector by end use for the year 2017.

Data from Environment and Climate Change Canada, 2017. National Inventory Report 1990–2016: Greenhouse Gas Sources and Sinks in Canada, Ottawa.

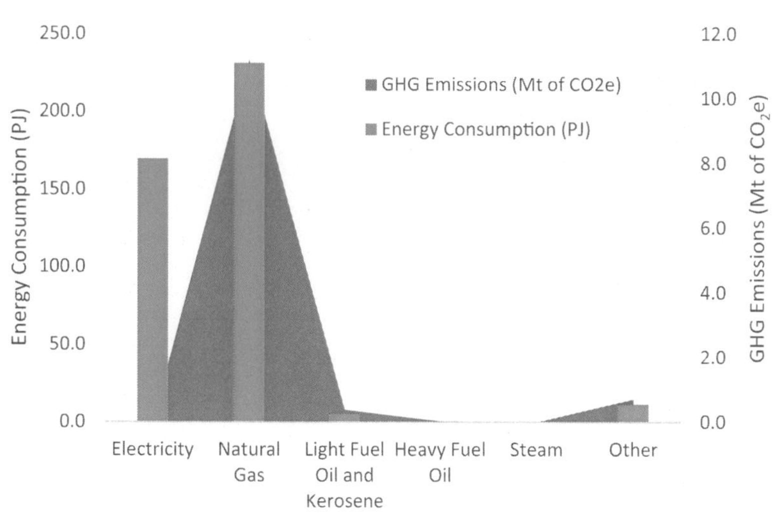

FIGURE 7.15

Energy consumption and GHG emissions in the commercial sector by energy source for the year 2017.

Data from Statistics Canada, 2017. Report on Energy Supply and Demand in Canada, 1990–2016, CANSIM (Table 128-0016), Ottawa.

Moreover, the sources of energy to meet the demand for this sector influence the environmental impact of this sector. Natural gas accounts for 77% of the energy supplied to this sector. Fig. 7.15 shows the energy consumption and GHG emissions per energy source in this sector. The energy commodities associated with the commercial sector revolve around electricity consumption as well as heating services such as space heating, water heating, and drying. Lighting and auxiliary equipment are usually powered using electricity. The electricity grid mix differs from region to another. Usually, nuclear energy, hydropower, and fossil fuels make up the based load for electric generation; whereas, renewables such as solar, wind, and geothermal are used for fluctuating and peak demands.

Other sources represent coal and propane, and their environmental impact is minimal. Furthermore, heavy fuel oil and steam are insignificant. Electricity and natural gas are the two primary sources for this sector. The environmental impact of electricity for this sector is excluded, leaving natural gas scoring the highest in GHG emissions.

The commercial sector houses have various activities from educational services to wholesale trade activities. Fig. 7.16 shows the energy consumption and GHG emissions per activity type. Offices include activities related to finance and insurance; real estate and rental and leasing; professional, scientific and technical services; public administration; and others.

Office is the top activity type in this sector requiring energy. This is in line with the previous observations about the primary commodity of space heating and the

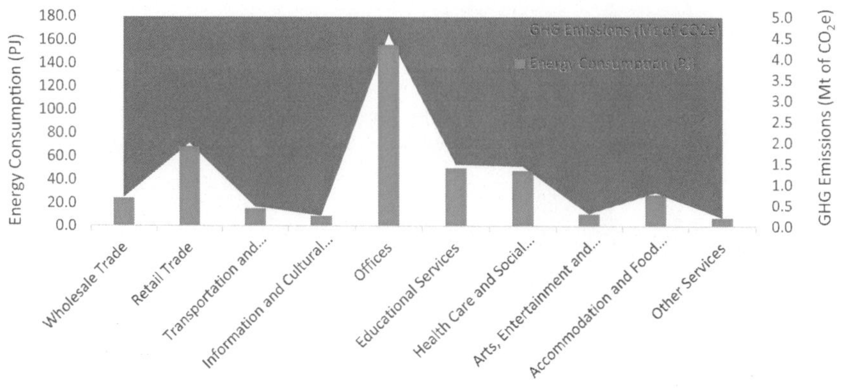

FIGURE 7.16

Energy consumption and GHG emissions in the commercial sector by activity type for the year 2017.

Data from Natural Resources Canada, 2017.

main source of energy, natural gas. It is also reasonable that it accounts for the highest GHG emissions in this sector. Electricity is excluded from the GHG emission calculations. Retail trade, educational services, healthcare, and social assistance activities are relevant in their energy consumption and GHG emissions.

The summary of the environmental impact section is presented here. These highlights reflect the main findings from this section.

- The industrial sector demands the highest amount of energy in Ontario followed by the transportation sector.
- Natural gas is a significant contributor to industrial, residential, and commercial sectors, yielding substantial GHG emissions and an adverse environmental impact.
- Motor gasoline and diesel are the primary energy sources for the transportation sector, both of which are fossil-based fuels with undesirable environmental impact.
- The iron and steel, petroleum refining, and pulp and paper industries constitute the top three industries requiring energy in this sector, consuming approximately 342 PJ, which is almost 70% of the total consumed energy in this sector
- Passenger light trucks and heavy trucks constitute most of the energy demand in the transportation sector, accounting for approximately 70% of the total energy demand. They also account for 41% of the total GHG emissions while cars are responsible for 27%.
- Space heating and water heating are the two main commodities in the residential sector requiring energy. Their energy demands make up more than 90% of the total energy demand. Natural gas alone accounts for 88% of all emissions in this sector.

- Single detached buildings are the primary building types in the residential sector requiring energy for space and water heating.
- Space heating is the primary commodity in the commercial sector that require energy as it constitutes more than 70% of the total energy demand in this sector. This commodity is also associated with natural gas and the highest GHG emission ratio. Natural gas makes up 77% of the energy supplied to this sector.

7.1.4 Economic performance assessment

As natural gas and electricity are the two primary sources of energy fueling various sectors in Ontario, the economic performance of these two sources will be assessed thoroughly in this section to understand the economic performance of Ontario's energy for the year 2017.

7.1.4.1 Natural gas

Ontario's natural gas is supplied by three main distributors, namely: Enbridge, Union Gas, and Natural Resource Gas (NRG). The economic performance of these three distributors reflects on the economic portfolio for natural gas in Ontario. Fig. 7.17 shows the total assets and liabilities for each distributor for the year 2017.

The general trend for all distributors reflects a positive financial performance with assets being more than liabilities. The total assets are \$17.8 billion while the total liabilities are \$12.3 billion with a 0.73 ratio between these two variables. It is also evident that Enbridge accounts for 78% of the total assets while Union Gas accounts for 42%. NRG's contribution is very insignificant. Current assets include gas inventories, accounts receivable, and cash accounts. Noncurrent assets

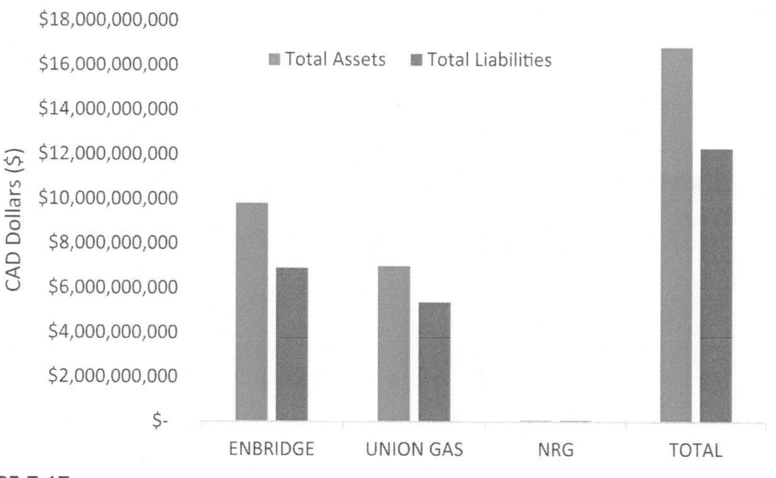

FIGURE 7.17

Total assets and liabilities of the natural gas distributors in Ontario for the year 2017.

Data from OEB, 2016.

include property, plant, equipment, and long-term investments. Fig. 7.18 illustrates the revenues and expenses for these distributors in 2017.

Although Enbridge's revenue is the highest, its expenses are also significant, resulting in a near break-even point. In fact, the net operating income incurred by Union Gas is more than that of Enbridge's. Furthermore, the total revenues from this energy source are approximately $7.4 billion, whereas the expenses are approximately $7.0 billion. The revenues incurred mainly come from the operating revenues, whereas the expenses are due to gas cost, operating, and maintenance in addition to interest. Furthermore, Fig. 7.19 shows the total number of customers

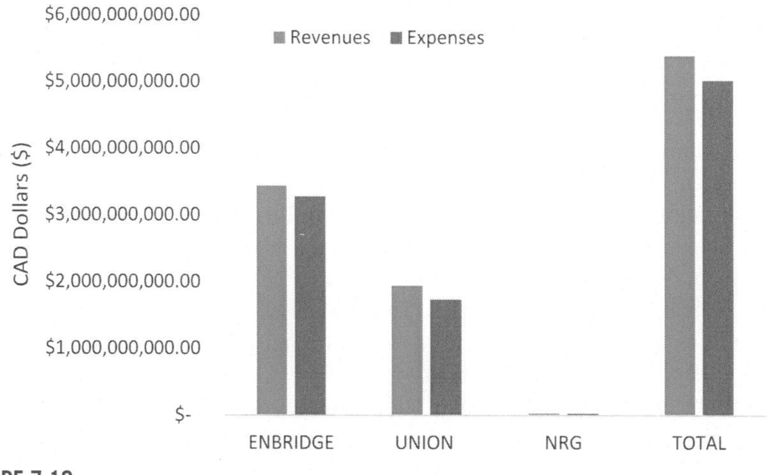

FIGURE 7.18

Total revenues and expenses of the natural gas distributors in Ontario for the year 2017.

Data from OEB, 2016.

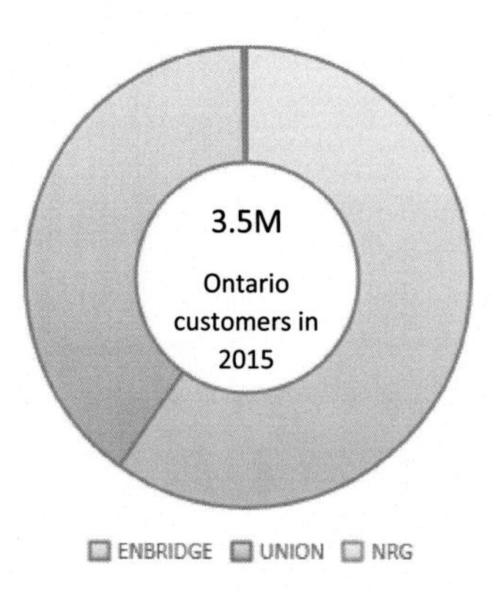

FIGURE 7.19

Total number of customers in Ontario for the year 2017.

Data from OEB, 2016.

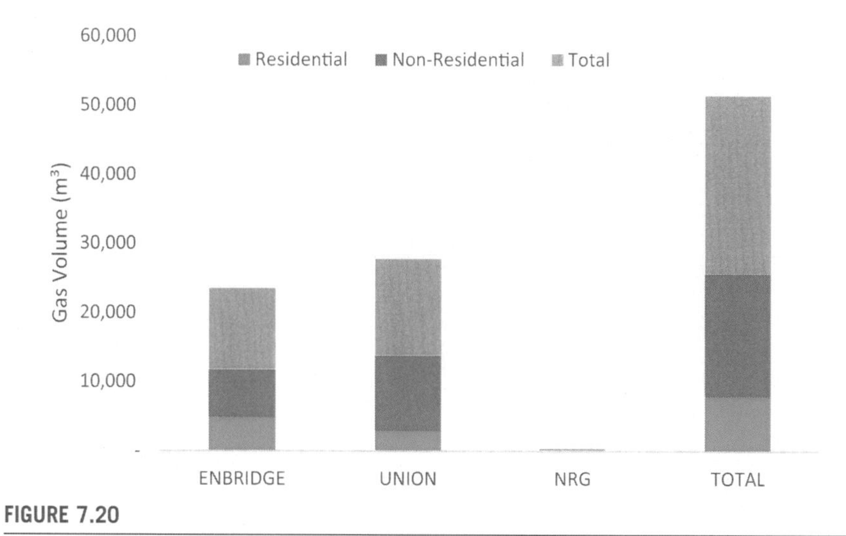

FIGURE 7.20

Natural gas supply to residential and nonresidential applications in Ontario for the year 2017.

Data from OEB, 2016.

in 2017. Enbridge supplies to approximately 70% of all Ontario customers, whereas Union Gas supplies to approximately 40%. The total number of customers includes system gas customers and direct purchase customers of marketers licensed by the OEB. The total number of customers adds up to approximately 3.7 million customers across Ontario. These customers vary as some are residential and some are nonresidential customers. In fact, approximately 70% of the total nature gas volume is supplied to nonresidential applications, whereas the remaining 30% are supplied for residential use. Fig. 7.20 shows the amount of natural gas supplied to each sector by each distributor.

It is evident that Union Gas is the main supplier of natural gas to nonresidential applications including industrial, commercial, and transportation sectors. It is indeed responsible for 71% of all natural gas supplied to nonresidential use. On the other hand, Enbridge balances supply between residential and nonresidential applications. Union Gas supplies relatively higher natural gas volume than Enbridge accounting for 74% of the total supplied natural gas in Ontario. Moreover, Fig. 7.21 illustrates the ratio between residential and nonresidential supply of natural gas. As mentioned earlier, natural gas primarily supplies nonresidential applications whereas residential use of natural gas reaches up to 30% of all supplied natural gas.

On the other hand, economic impacts of reducing natural gas and electricity use in Ontario have been analyzed by Stokes Economic Consulting Inc. (2013). They claim that reducing both natural gas and electricity use in Ontario by 27% over the period to 2027 increases Ontario's real GDP by $3.7 billion and employment by over 27, 000. They also claim that it substantially reduces the investment in

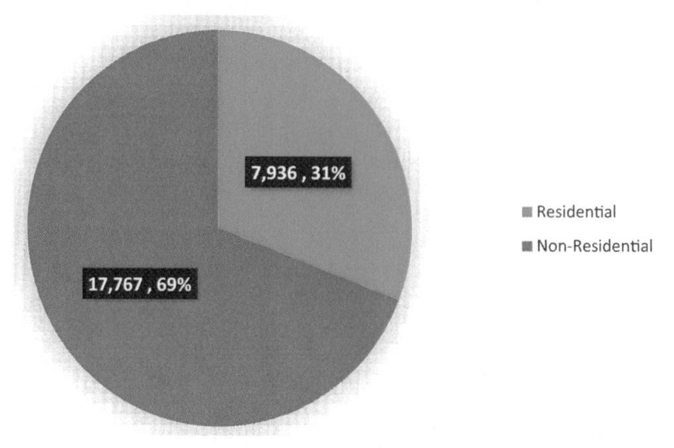

FIGURE 7.21

Natural gas volume (m^3) supplied to residential and nonresidential applications in Ontario for the year 2017.

Data from OEB, 2016.

the electric power generation industry as well as improving Ontario's trade balance by significantly reducing imports of natural gas. Furthermore, they claim that the federal budget balance can be improved by $1 billion by 2027 alongside a $982 million improvement in the Ontario government balance (SEC, 2013).

7.1.4.2 Electricity

Ontario's electricity is unique as it is supplied by a diverse mix of nuclear, hydro, and renewable energy sources. The overall cost of electricity has increased significantly since 2007 due to phase out of coal generation and the expansion of natural gas capacity (Fremeth et al., 2017). Furthermore, energy impacts Ontario's economy significantly as electricity alone is a $17 billion annual industry accounting for 8% of Canada's GDP. Ontario Energy Board (OEB) is Ontario's independent energy regulator who establishes electricity and natural gas rates. They also oversee energy companies to ensure that they comply with provincial rules. They track utilities performance and monitor the wholesale electricity market. On the other hand, the independent electricity service operator (IESO) is the body that balances the supply and demand of electricity on a second-by-second basis. It is responsible for directing electricity flow across Ontario's high voltage transmission lines to make it available for consumers. The electricity is distributed to consumers across Ontario through 71 local distribution companies (LDCs). These LDCs have specific local jurisdictions, and they exist to provide competitive and more efficient electricity supply. The total liabilities and assets for Ontario's LDCs are demonstrated in Fig. 7.22. More assets than liabilities reflect a positive and healthy performance for these LDCs collectively. Assets from this source of energy are incurred from receivables, property

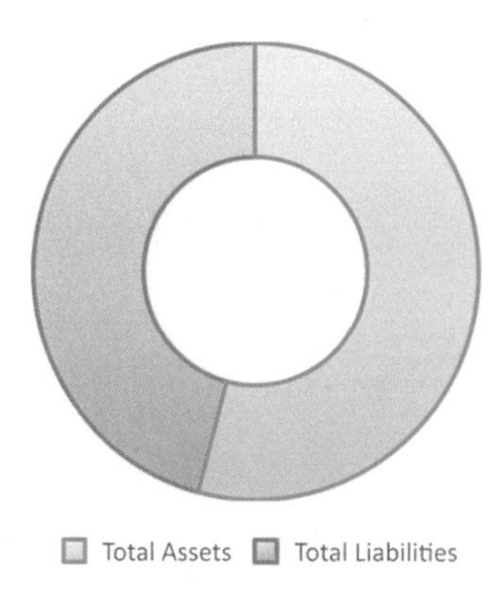

FIGURE 7.22

Total assets and liabilities of the electricity LDCs in Ontario for the year 2017.

Data from OEB, 2016.

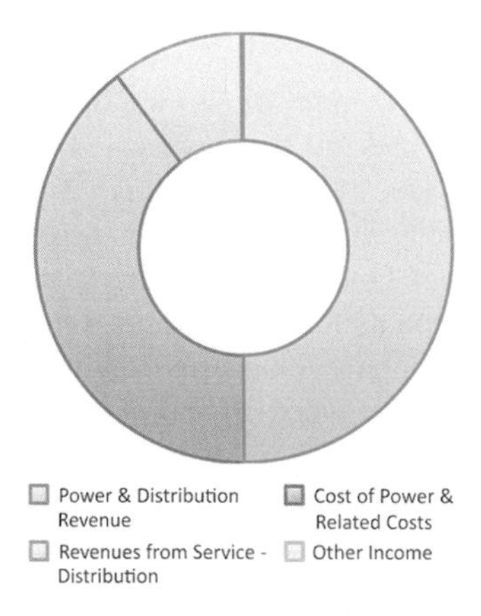

FIGURE 7.23

Revenue sources from electricity in Ontario for the year 2017.

Data from OEB, 2016.

plants, equipment, regulatory, and other assets. On the other hand, liabilities include accounts payable, intercompany payables, long-term debt, and employee future benefits. Fig. 7.23 illustrates the income further demonstrating that power and distribution revenue account for 70% of all incurred revenue. The second major source of revenue is the cost of power and related costs. Both of these sources account for 90% of all incurred revenue in Ontario's electricity.

On the other hand, expenses sustained from electricity in Ontario are primarily due to administration, depreciation, and amortization as shown in Fig. 7.24. In fact, these primary expenses account for 70% of the total expenses in Ontario's electricity. Financing, maintenance, and operating also incur notable amount on the expenses sheet.

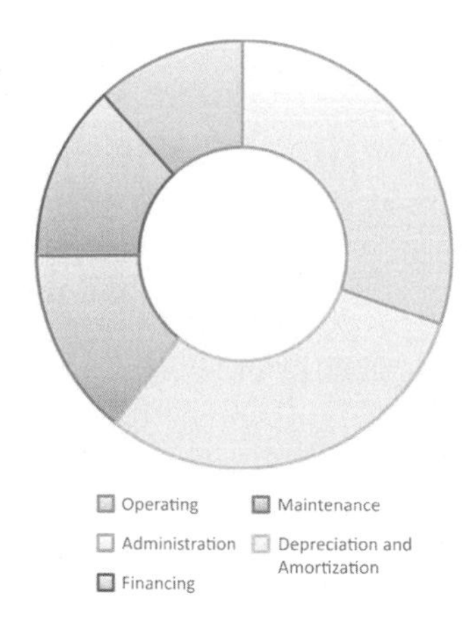

FIGURE 7.24

Expenses sustained from electricity in Ontario for the year 2017.

Data from OEB, 2016.

More information about the nature of service that electricity provides is presented in Table 7.1. This table presents the background about the customers, amount of electricity delivered, and key additions for this year. The top five LDCs were selected further to assess their economic performance. These five LDCs account for 87% of the total assets of Ontario's electricity. The top LDC namely Hydro One Networks Inc. alone accounts for 70% of all assets and 71% of all liabilities.

This makes performance enhancements and optimization associated with this single LDC very impactful on the overall economic performance of Ontario's electricity. Fig. 7.25 shows the total assets and liabilities of the top five LDCs.

All these LDCs have higher assets than liabilities reflecting positive performance. However, the difference between the two values is relatively small, indicating that enhancement can be applied to yield better economic performance. As Hydro One Networks Inc. is the major supplier of electricity in Ontario, further analysis about it is presented in Table 7.2.

Residential customers are the main customers for Hydro One Networks in electric consumption and revenue. General Service and Large Users score second in electric consumption. They are very similar in revenue to the general service customer of less than 70 kW.

This means that larger users consume more, yet the revenue incurred is not proportional compared to other types of consumers. Fig. 7.26 shows the various trends extracted from Table 7.2. The summary of the economic performance section is presented here. These highlights reflect the main findings from this section.

- Enbridge accounts for 78% of the total assets in Natural Gas, whereas Union Gas accounts for 42%.
- The total revenues from Natural Gas are approximately $7.4 billion, whereas the expenses are approximately $7.0 billion.

Table 7.1 Key parameters highlighted from the electricity industry in Ontario for the year 2017.

Parameter	Unit
Total customers	7,074,739
Residential customers	4,774,837
General service <70 kW customers	434,999
General service (70–4999 kW) customers	74,297
Large user (>7000 kW) customers	124
Total service area (km^2)	991,733
% Rural	99%
% Urban	1%
Total km of line	199,477
Overhead km of line	173,377
Underground km of line	47,099
Total kWh supplied	1.24729E+11
Total kWh delivered (excluding losses)	1.19901E+11
Total kWh delivered on long-term load transfer	88,704,709.72
Total distribution losses (kWh)	4,739,779,717
Gross capital additions for the year ($)	$ 2,230,339,204.73
High voltage capital additions for the year ($)	$ 807,133,010.47

Data from OEB, 2016.

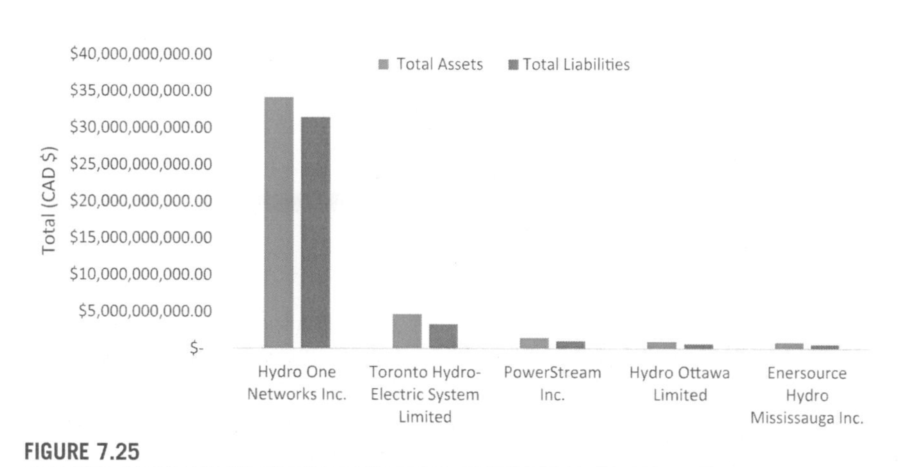

FIGURE 7.25

Total assets and liabilities of the top seven LDCs in Ontario for the year 2017.

Data from OEB, 2016.

Table 7.2 Key parameters associated with Hydro One Networks in 2017.

Parameter	Unit
Residential customers	
Number of customers	1,141,379
Metered kWh	13,077,791,987
Distribution revenue	$872,779,937
Metered kWh per customer	11,448
Distribution revenue per customer	$747
General service <70 kW customers	
Number of customers	107,270
Metered kWh	2,999,770,098
Distribution revenue	$170,974,842
Metered kWh per customer	27,977
Distribution revenue per customer	$1794
General service >70 kW, large user (>7000 kW), and subtransmission	
Number of GS >70 kW customers	7901
Number of large users	–
Number of subtransmission customers	487
Metered kWh	7,449,839,774
Distribution revenue	$172,774,777
Metered kWh per customer	749,797
Distribution revenue per customer	$19,407
Unmetered scattered load connections	
Number of connections	7497
Metered kWh	30,733,843
Distribution revenue	$3,747,821
Metered kWh per connection	7777
Distribution revenue per connection	$747

Data from OEB, 2016.

- Enbridge supplies to approximately 70% of all Ontario customers, whereas Union Gas supplies to approximately 40%. The total number of customers adds up to approximately 3.7 million customers across Ontario
- Approximately 70% of the total nature gas volume is supplied to nonresidential applications, whereas the remaining 30% are supplied for residential use
- Union Gas supplies relatively higher natural gas volume than Enbridge accounting for 74% of the total supplied natural gas in Ontario.
- Electricity is distributed across Ontario through 71 LDCs.

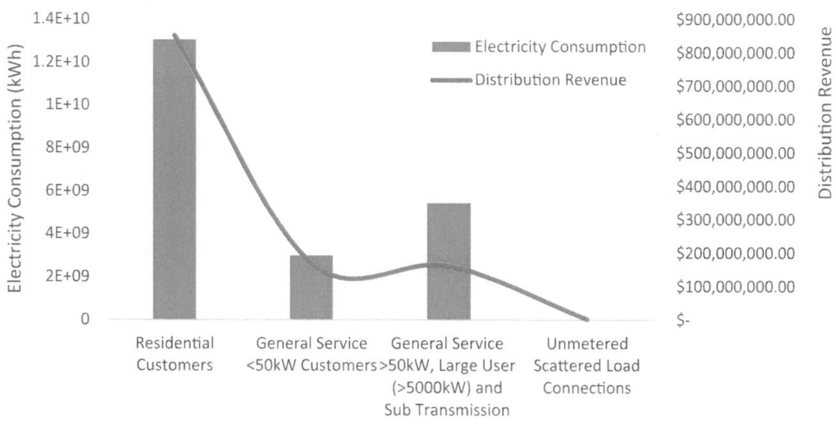

FIGURE 7.26

Hydro One Networks consumers' electricity consumption and distributed.

Data from OEB, 2016.

- Power and distribution revenue accounts for 70% of all incurred revenue. The second major source of revenue is the cost of power and related costs. Both of these sources account for 90% of all incurred revenue in Ontario's electricity.
- The top five LDCs account for 87% of the total assets of Ontario's electricity. The top LDC namely Hydro One Networks Inc. alone accounts for 70% of all electricity assets and 71% of all liabilities.
- Residential customers are the main customers for Hydro One Networks in electric consumption and revenue.

7.1.5 Social impact assessment

The social influence of the energy sector in Ontario is explored in this section. The objective is to truly reflect the social impacts experienced from Ontario's present energy portfolio. A number of indicators are assessed namely job creation, social acceptance, public awareness, and perceived cost of electricity. Pirnia et al. (2011) analyzed the policy implications of Ontario's FITs on overall societal welfare. They suggest that existing FIT tariffs would have a large negative impact on consumer welfare, with an overall net loss on total social welfare if they were unbounded. Similar to the economic performance, the social aspect analysis focuses on the two main energy sources in Ontario: natural gas and electricity. Green supply chain management is highlighted in the research of Govindan (2017) who concentrated on green sourcing as a mean to achieve sustainability management and conservation of resources. As for electricity, nuclear and hydropower are assessed in specific as they make up 81% of Ontario's electricity mix (Ministry of Energy, 2017).

7.1.5.1 Job creation

The energy sector in Ontario employees around 97,000 people directly and indirectly. Furthermore, Ontario attracted more than $17 billion in private sector investments in the energy sector in 2017 (Ministry of Energy, 2017).

The main sources of energy, where job creation is assessed include natural gas, nuclear, and hydropower. These sources are the vital energy sources that fuel Ontario. Table 7.3 shows these energy sources and their OECD employment factors used in the 2017 global analysis conducted by Rutovitz et al. (2015).

Nuclear and small hydro projects yield longer job years per MW during construction and installation phases. Furthermore, small hydro project results in the highest job years per MW during manufacturing phases. As for operation and maintenance, nuclear power results in 0.7 jobs for every MW, whereas gas yields 0.14 jobs for every MW. Furthermore, gas is the least energy source for job creation in all phases of the project.

After averaging the global employment factors for each source based on the data presented in Table 7.3, the capacity for each energy source was assessed to assess how many jobs each energy source provides to Ontario. Fig. 7.27 shows the job creation projection in Ontario.

Nuclear and hydropower compete in providing jobs to Ontarians. It seems that hydropower is attractive socially as it provides longer job opportunities. Furthermore, nuclear power is also a favorable choice for energy as it provides a competitive number of jobs. Realistically, these jobs may be repetitive as these jobs reflect the work required throughout the life cycle of each project starting from construction, installation, manufacturing throughout operation, and maintenance.

7.1.5.2 Social acceptance and public awareness

Most studies of social acceptance of energy projects are associated with renewable energy sources, wind energy in specific. Not much research has been done on Ontario's public acceptance on natural gas, nuclear, hydro, or other major sources of energy. In fact, such studies are scarce and done distinctly from one another. For example, the National Nuclear Attitude Survey claims that most Ontarians are in support of nuclear power with greater support from men than women (Innovative

Table 7.3 Main energy sources and their global employment factors in 2017.

Energy source	Construction/Installation (job years/MW)	Manufacturing (job years/MW)	O&M (jobs/MW)
Gas	1.3	0.93	0.14
Nuclear	11.8	1.3	0.7
Hydro-large	7.4	3.7	0.2
Hydro-small	17.8	10.9	4.9

Data from Rutovitz et al., 2015.

Research Group, 2012). However, it is evident that Ontario energy plan ensures the gradual engagement of the public and specific communities to raise awareness and move in the direction of sustainable energy growth with social acceptance in mind. However, the plan tends to confine the social aspect by attending to the First Nation and Metis communities. On the other hand, Ontario did enhance the social aspect slightly when formulating the 2017 Long-Term Energy Plan by establishing public engagement sessions (Ministry of Energy, 2017). These sessions included 17 public open houses, attended by hundreds of Ontarians, 770 stakeholders from the energy sectors, the business community, and various municipalities. These stakeholders discussed energy delivery, innovation, energy supply, conservation, and efficiency along with energy prices. Furthermore, 17 indigenous engagement sessions were held across 100 indigenous communities and organizations. Lastly, 2377 online submissions were received surrounding energy themes such as delivery, supply, prices, innovation, conservation, and indigenous engagement. The summary of these submissions is illustrated in Fig. 7.28.

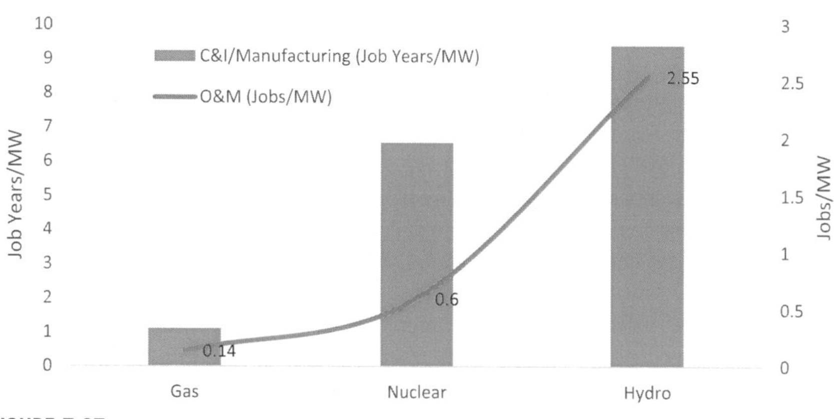

FIGURE 7.27

Job creation relative to each energy source for Ontario.

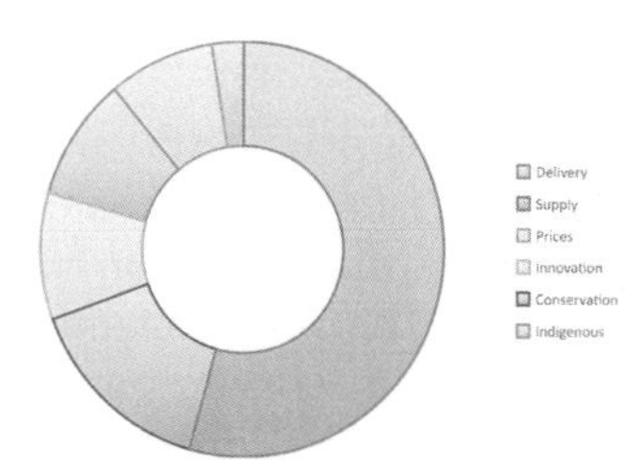

FIGURE 7.28

Summary of social engagement results for Ontario's energy sector in 2017.

Data from Ministry of Energy, 2017.

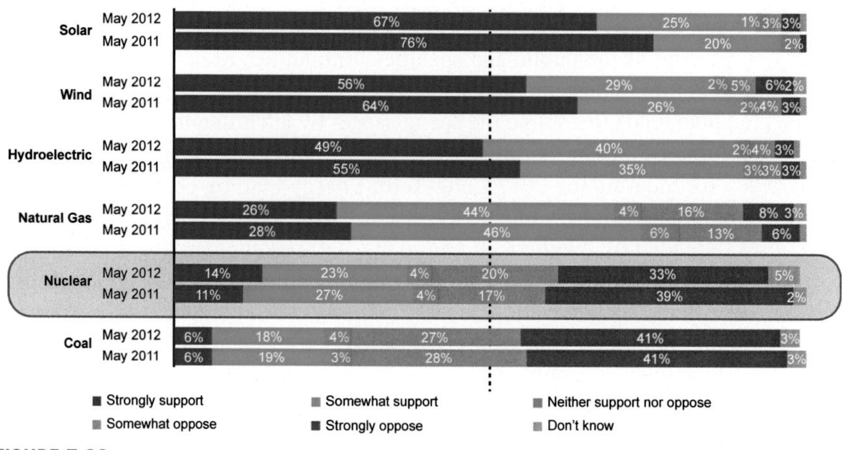

FIGURE 7.29

Survey results of support and opposition of various forms of electricity generation in Canada.

Figure from Innovative Research Group, 2012.

Supply of energy is a major theme of interest to Ontarians followed by conservation of energy, innovation, and delivery. Interestingly, prices did not receive many submissions from the public, which could be interpreted that the public is pleased with the cost of energy. Furthermore, energy supply reflects that delivery of energy services needs to be enhanced more. Therefore, for Ontario, energy sources and systems may be satisfactory while energy service and storage solutions need to be enhanced to achieve higher social acceptance and thus higher energy sustainability.

In addition, the National Nuclear Attitude Survey published results of total support for various electricity generation platforms. Fig. 7.29 presents a summary of these results. Clearly, there is strong support for renewable energy sources such as solar, wind, and hydroelectric sources. Natural gas has slightly declining support with most supporters partially supporting it. Nuclear and coal have evident opposition in Canada with some supporters for both sources.

7.1.5.3 Perceived cost of electricity

The National Nuclear Attitude Survey presents the perceived cost of electricity through a public survey in 2012. The public rated each energy source used in electricity production keeping in mind the overall cost from building, a lifetime of generating and finally decommissioning. The results of this survey are presented in Fig. 7.30. The responses are very dynamic as many sources are rated as very expensive or somewhat expensive. Furthermore, solar power has the highest ratings of all that it is somewhat inexpensive. All sources had 4%–7% of neutral participants.

The summary of the social aspect is presented here. These highlights reflect the main findings from this section.

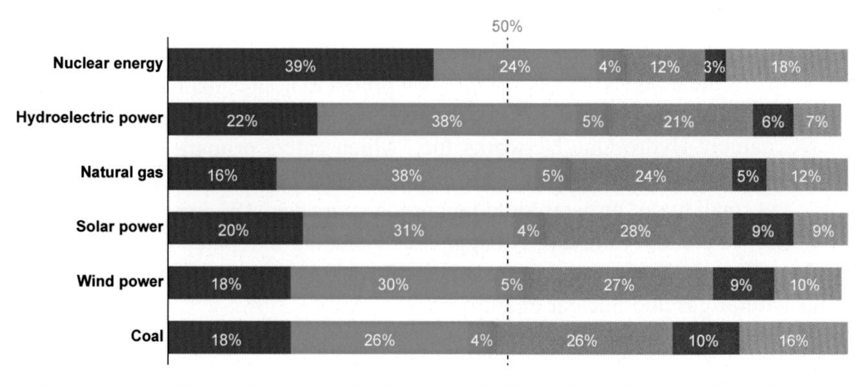

FIGURE 7.30

National survey results of public rating of energy cost per source.

Figure from Innovative Research Group, 2012.

- The social aspect analysis focuses on the two main energy sources in Ontario: natural gas and electricity.
- Nuclear and small hydro projects yield longer job years per MW during construction and installation phases.
- Nuclear and hydropower compete in providing jobs to Ontarians while Gas is not favorable.
- Ontario did enhance the social aspect slightly when formulating the 2017 Long-Term Energy Plan by establishing public engagement sessions.
- 2377 online submissions were received surrounding energy themes such as delivery, supply, prices, innovation, conservation, and indigenous engagement.
- There is strong support for renewable energy sources such as solar, wind, and hydroelectric sources. Natural gas has slightly declining support with most supporters partially supporting it. Nuclear and coal have evident opposition in Canada with some supporters for both sources.

7.1.6 Closing remarks

In conclusion, natural gas and electricity are the major energy sources in Ontario throughout all sectors. The industrial sector however dominates the energy demand by using fossil-based fuels, especially natural gas. Moreover, the transportation sector relies significantly on diesel and gasoline for fuel. The industrial and transportation sectors combined constitute most Ontario's energy demands. Heavily fueled by natural, these sectors contribute to significant GHG emissions. Specifically, the iron and steel industry is one of the major industries consuming energy in Ontario. As for transportation, passenger light trucks and heavy trucks constitute most of the energy demand in the transportation sector, accounting for approximately 70% of the total energy demand in this sector. Space and water heating are the main commodities demanding energy in the residential sector. Economically, natural gas in Ontario yields $7.4 billion in revenues and $7.0 billion in expenses. Approximately 70% of the total nature gas volume is supplied to nonresidential applications. The top five LDCs account for 87% of the total assets of Ontario's electricity. The top LDC namely Hydro One Networks Inc., which caters primarily to residential customers, accounts for 70% of all electricity assets and 71% of all liabilities. Lastly, nuclear and small hydro projects yield longer job years per MW during construction and installation phases. Community engagement is being integrated in Ontario's energy plan. There is vibrant support for renewable energy sources and declining support for natural gas and nuclear energy.

7.2 Case study 2: sustainability assessment of an integrated net zero energy house

In this study, a net zero energy house is considered and modeled using solar PV and geothermal heat pump. The system is assessed for sustainability and energetic and exergetic efficiencies. The solar system considered yields an electricity production of 71.4 kW with exergy efficiency of 17% under atmospheric conditions. The geothermal heat pump has a coefficient of performance of 4.9 and an exergetic coefficient of performance of 2.1. The sustainability index of this system is 0.72 using the hierarchist aggregation method and the weighted geometric mean. Furthermore, the effect of various refrigerants on the thermodynamic performance has been investigated.

7.2.1 Introduction

Energy management and efficiency enhancements of energy systems have been key strategies in reducing cost and environmental impact. Implementing such practices and technologies, energy consumption of buildings can be lowered by up to 37% (Ayoub, 2013). Assessment for sustainability of energy systems is significant as it provides a detailed layout of all parameters pertaining energy consumption. A comprehensive sustainability assessment allows analyzing all system-related parameters including environmental, economic, social, and technology impacts along with evaluating thermodynamic-based properties. Santoyo-Castelazo and Azapagic (2014) included three main indicators to assess the economic category of their sustainability model. They used capital costs, total annualized costs, and levelized costs to assess the sustainability of energy systems. They also investigated the social category more comprehensively than other studies by investigating security and diversity of supply of energy, public acceptability, health and safety, and intergenerational issues. Furthermore, the concept of net zero energy houses are widely spreading as their long-term benefits are attractive to national plans of many countries worldwide (Ayoub, 2013). For example, the state of California has a policy goal that all new low-rise residences be net zero houses by the year 2020 along with all commercial and high-rise buildings by 2030. Boza-Kiss et al. (2013) evaluated policy instruments to foster energy efficiency for the sustainable transformation of buildings. They found that all policy instruments have the potential to generate economic benefit. Furthermore, regulations can be cost-effective in various environments. In addition, public leadership procurement programs have hidden impacts under other instruments. Their data analysis suggests that regulations associated with product energy performance standards have the ability to have the largest lifetime energy saving impacts. Furthermore, Kahn et al. (2013) made it evident that it is an intrinsic challenge to develop sustainable energy solutions with an increasing demand for energy services. They also highlight that environment should be enabled to facilitate rapid transformation of current energy systems. On another note, Modarres (2017) investigated the relationship between commuting, energy consumption, and the challenges associated with urban development. They analyzed socioeconomic patterns and their impact on energy consumption and thus sustainable development. Furthermore, Yi et al. (2017) investigated a net zero energy building from an ecological perspective and assessed the sustainability based on emergy theory. Their results show that net zero energy buildings use greater nonrenewable emergy to seek a zero-energy budget. Moreover, they claim that sustainable buildings tend to maximize power and not efficiency. In addition, Ulubeyli and Kazanci (2018) developed a holistic sustainability assessment of green buildings. Their model included environmental, legal, political, economic, technological, and social factors. Their conclusion indicates that technological and environmental factors were the most influential factors when it comes to sustainability of buildings.

Moreover, Polzin et al. (2017) claim that the current financial system is not conducive to an innovation-led energy transition. They indicate that there is a diverse investment demand to make the transition to a more sustainable energy system. They conclude that higher diversity and resilience in financial markets is instrumental to facilitate the transition to clean energy in our current economy. Lastly, Tirado et al. (2018) tried to address a critical question whether smart home technologies address key energy challenges in households or not. They claim that real-life evidence of the impact of smart home technologies on everyday life of households is rare. Smart home technologies enable novel ways to use and manage energy in a domestic sphere. In fact, they argue that smart home technologies may reinforce unsustainable energy consumption patterns in the residential arena and that they are not readily available.

There are technically evident gaps and challenges associated with the concept of a net zero energy house. Although net zero energy houses require a large capital cost, a quantitative economic assessment remains necessary. Moreover, environmental and social impacts of net zero energy houses also need to be studied in depth.

The latest technologies and telecommunications techniques in net zero energy house applications have been reviewed by AlFaris et al. (2017) who concluded that the renewable energy systems with smart houses are the most cost-effective options. They also claim that the energy performance of such systems is 37% better than ASHRAE standards when integrating renewable energy in smart houses.

Furthermore, the energy performance of net-zero and near net-zero energy houses were analyzed in New England and 7 out of 10 houses achieved the net-zero standard or better (Thomas and Duffy, 2013). They also concluded that behavior of occupants, extra equipment, and mechanical problems affected individual energy consumption. Payback time along with other life cycle analysis of a net zero energy house is presented by Leckner and Zmeureanu (2011). They claim that the energy payback time from solar in this house is between 8.4 and 8.7 years. They also concluded that solar energy conversion also helps in supplying at least 3.7 times more energy than the energy invested for manufacturing and shipping the system. Further financial analysis is presented by Delisle (2011) where they evaluated the potential impact of the price fluctuation on net zero energy houses construction costs. Their study also compared various combinations of energy efficiency parameters and the annual energy production.

Although net zero energy house is an attractive idea, which may seem more financially sound and technically feasible, its sustainability or performance is not yet examined in depth. This paper aims to fill this gap by analyzing a net zero energy house, assessing its performance thermodynamically as well as determine the sustainability índex associated with this idea. This idea of using the thermodynamic performance along with other major impact of energy systems to assess their sustainability index is novel. Both first and second laws of thermodynamics were proposed as the primary criteria of a proposed model to investigate technology development and long-term energy transition (Wang et al., 2017). In fact, the authors follow evolutionary forecasting scenarios and use the laws of thermodynamics to simulate the development of energy transformation technologies. This paper presents a comprehensive sustainability assessment of a net zero energy house by evaluating the energy system's thermodynamic performance along with other indicators such as economic, social, and environmental impacts. The specific objectives of this study are listed as follows:

- To analyze and assess the performance of a net zero energy house, which uses solar PV and geothermal heat pump.
- To develop a model and assess the sustainability level of the considered system using a comprehensive sustainability assessment model.
- To investigate how changing operating and environment conditions will affect the thermodynamic performance and sustainability of the applied system.

7.2.2 System description

The sustainability assessment is conducted on a residential house in Ontario, Canada. The house is a two-story, 177 m^2 house with an annual natural gas consumption of 37,771 kWh and an electric consumption of 7331 kWh annually. The house is located in Bowmanville, specifically at latitude 43.91 and longitude −78.78. The system used to achieve net zero energy comprises a grid-integrated, roof mounted solar PV system, backed up with battery storage. Although the solar PV provides

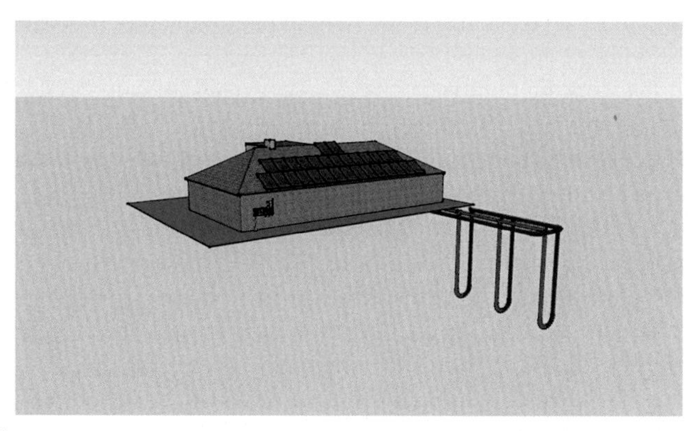

FIGURE 7.31

A 3D illustration of the roof mounted solar system and the geothermal ground vertical loops.

electricity, a ground-source heat pump is used for heating and cooling applications. Fig. 7.31 illustrates how both systems are connected to the house.

Solar PV panels are installed on a south-facing roof, with no shading that surrounds the roof. The panels provide electric power, which is harvested from the sun and converted from direct current to alternating current. Electricity generated through solar is an alternative to current electricity supplied by Ontario energy mix, which is 70% nuclear, 23% hydroelectricity, 10% from natural gas, and the remainder from other renewable sources (Ontario Ministry of Energy, 2017). Excess electricity is integrated back to the grid under the microFIT provincial incentive-based program. Furthermore, heating, cooling, and domestic hot water are supplied by a geothermal vertical loop that is connected with a heat pump system. The design is based on an Ontario residential house of which the electric and gas usage are used for this assessment.

7.2.2.1 Geothermal system

The heat pump is also modeled using the Engineering Equations Solver (EES) to analyze the system thermodynamically. The heat pump is designed to meet 90% of the heating demand for the building during the winter season. The demand is evaluated based on monthly utility bills. For this site, natural gas was used for both heating and for hot water. Balance equations for the different states in the system are calculated to understand, the energetic and exergetic performance of the system. The heat pump uses water glycol solution in the vertical ground loop and ammonia as a refrigerant in the vapor compression cycle for the heat pump, which heats the air for the site. The system is designed to meet the demand of a 7.7 kW load. The vapor compression cycle operates by compressing the working fluid, ammonia, in the compressor to a high pressure and temperature. The compressor for the actual system was assumed to have an isentropic efficiency of 87%. The working fluid (gaseous phase) is then fed into the condenser, which acts as a heat exchanger. Air is heated as the working fluid is condensed. A 70 kPa pressure drop is assumed in the condenser. The working fluid then leaves the condenser as a saturated liquid. From the condenser, the working fluid is throttled in an expansion valve to the evaporator pressure. The throttling process is modeled as an isenthalpic process. The evaporator acts as a heat exchanger transferring heat from the water glycol solution circulating then in the vertical ground loop to the refrigerant. The working fluid enters the compressor as a saturated vapor, repeating the cycle. Although Fig. 7.31 illustrates the connection of the geothermal to the house, Fig. 7.32 shows the heat pump system.

It is assumed that kinetic and potential energy interactions are negligible. In addition, each component is analyzed as a control volume and at a steady state. The sequence of these equations follows the schematic sketch presented in Fig. 7.32. The compressor is labeled as state number 1 followed by the condenser, the expansion valve, and finally the evaporator as state number 4.

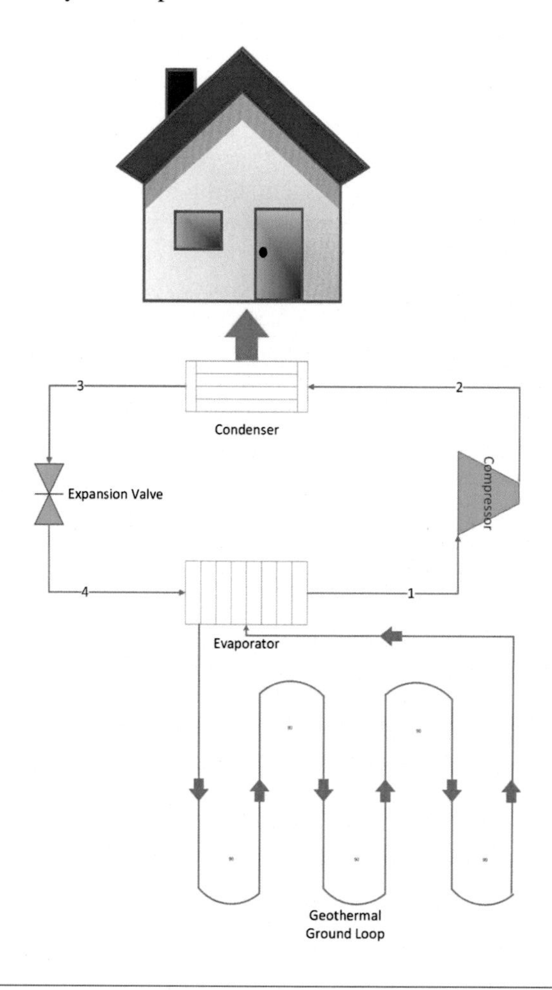

FIGURE 7.32

Sketch of the connection between the geothermal ground loop and the heat pump.

For the adiabatic compressor, one can write the thermodynamic balance equations as follows:

$$\text{Mass Balance Equation (MBE):}\quad \dot{m}_1 = \dot{m}_2 \tag{7.1}$$

$$\text{Energy Balance Equation (EBE):}\quad \dot{m}_1 h_1 + \dot{W}_{\text{comp}} = \dot{m}_2 h_2 \tag{7.2}$$

$$\text{Entropy Balance Equation (EnBE):}\quad \dot{m}_1 s_1 + \dot{S}_{\text{gen}} = \dot{m}_2 s_2 \tag{7.3}$$

$$\text{Exergy Balance Equation (ExBE):}\quad \dot{m}_1 ex_1 + \dot{W}_{\text{comp}} = \dot{m}_2 ex_2 + \dot{E}x_{\text{dest}} \tag{7.4}$$

For the condenser, one can write the thermodynamic balance equations as follows:

$$\text{Mass Balance Equation (MBE):}\quad \dot{m}_2 = \dot{m}_3 \tag{7.5}$$

$$\text{Energy Balance Equation (EBE):}\quad \dot{m}_2 h_2 = \dot{m}_3 h_3 + \dot{Q}_{\text{out}} \tag{7.6}$$

$$\text{Entropy Balance Equation (EnBE):}\quad \dot{m}_2 s_2 + \dot{S}_{\text{gen}} = \dot{m}_3 s_3 + \frac{\dot{Q}_{\text{out}}}{T_0} \tag{7.7}$$

$$\text{Exergy Balance Equation (ExBE):}\quad \dot{m}_2 ex_2 = \dot{m}_3 ex_3 + \dot{Q}_{out}\left(1 - \frac{T_0}{T_0}\right) + \dot{E}x_{\text{dest}} \tag{7.8}$$

For the expansion valve, one can write the thermodynamic balance equations as follows:

$$\text{Mass Balance Equation (MBE):}\quad \dot{m}_3 = \dot{m}_4 \tag{7.9}$$

$$\text{Energy Balance Equation (EBE):}\quad \dot{m}_3 h_3 = \dot{m}_4 h_4 \tag{7.10}$$

$$\text{Entropy Balance Equation (EnBE):}\quad \dot{m}_3 s_3 + \dot{S}_{\text{gen}} = \dot{m}_4 s_4 \tag{7.11}$$

$$\text{Exergy Balance Equation (ExBE):}\quad \dot{m}_3 ex_3 = \dot{m}_4 ex_4 + \dot{E}x_{\text{dest}} \tag{7.12}$$

For the evaporator, one can write the thermodynamic balance equations as follows:

$$\text{Mass Balance Equation (MBE):}\quad \dot{m}_4 = \dot{m}_1 \tag{7.13}$$

$$\text{Energy Balance Equation (EBE):}\quad \dot{m}_4 h_4 + \dot{Q}_{\text{in}} = \dot{m}_1 h_1 \tag{7.14}$$

$$\text{Entropy Balance Equation (EnBE):}\quad \dot{m}_4 s_4 + \frac{\dot{Q}_{\text{in}}}{T_{\text{space}}} + \dot{S}_{\text{gen}} = \dot{m}_1 s_1 \tag{7.15}$$

$$\text{Exergy Balance Equation (ExBE):}\quad \dot{m}_4 ex_4 + \dot{Q}_{\text{in}}\left(1 - \frac{T_0}{T_{\text{space}}}\right) = \dot{m}_1 ex_1 + \dot{E}x_{\text{dest}} \tag{7.16}$$

where the process is isenthalpic and entropy at state 4 can be found using enthalpy and pressure parameters.

7.2.2.2 Solar photovoltaic system

The electricity generated from the PV systems is used for the overall electric applications of the house as well as to provide the work input for the geothermal heat pump. Additional electricity is fed back to the grid. Due to the intermittent nature of solar energy and seasonal availability, electricity would be required from the grid during nights as well as other periods of low production.

The PV system is composed of 30 panels of 70-cell monocrystalline 300 W modules. The modules have a degradation rate of 0.77% annually and an initial degradation of 2%. It is assumed that all panels fit on the south-facing roof and that the roof is free of any shading surrounding it. Other mechanical and electrical properties of the panels used are presented in Table 7.4. The solar and geothermal heat pumps are modeled using the EES to analyze the system thermodynamically.

The heat pump coefficient of performance is calculated as follows:

$$\mathrm{COP_{en}} = \frac{\dot{Q}_\mathrm{H}}{\dot{W}_\mathrm{C}} \tag{7.17}$$

where $\dot{Q}_\mathrm{H}$is the heat output of the condenser, and $\dot{W}_\mathrm{C}$ is work output from the compressor. The energetic performance coefficient $\mathrm{COP_{en}}$ is determined by dividing

Table 7.4 Mechanical and electrical specifications of the module considered under nominal operating conditions.

Parameter	Specification
Cells	7 × 10
Cell type	Monocrystalline/P-Type
Cell dimensions (L × W × H)	171.7 × 171.7 mm/17.27 cm
Front load	7000 Pa
Rear load	7400 Pa
Weight	18.0 kg
Connector type	MC4, JM701A
Maximum power (Pmax)	220 W
MPP voltage (Vmpp)	29.1 V
MPP current (Impp)	7.77 A
Open circuit voltage (Voc)	37.0 V
Short circuit current (Isc)	8.10 A
Module efficiency	17.7%
Operating temperature	−40 ~ +90

the first term by the second. Furthermore, the exergetic performance coefficient COP_{ex} is calculated using the following equation:

$$COP_{ex} = \frac{\dot{E}x^{QH}}{\dot{W}_C} \tag{7.18}$$

where $\dot{E}x^{QH} = \left(1 - \frac{T_0}{T_H}\right) \times \dot{Q}_H$ is the total exergy of the condenser's heat output.

7.2.3 Sustainability assessment

These two renewable energy systems are assessed from a sustainability perspective to gauge the ratio of sustainability they achieve compared to other systems. The sustainability assessment model used is a comprehensive method that integrates various variables together to derive a dimensionless sustainability index. The assessment model takes into consideration, environmental, social, and economic impacts along with thermodynamic energy and exergy performances, technology, educational, and sizing indexes. Fig. 7.33 demonstrates the assessment model used for this study. The energy index accounts for the energy impact on other indexes as the environmental

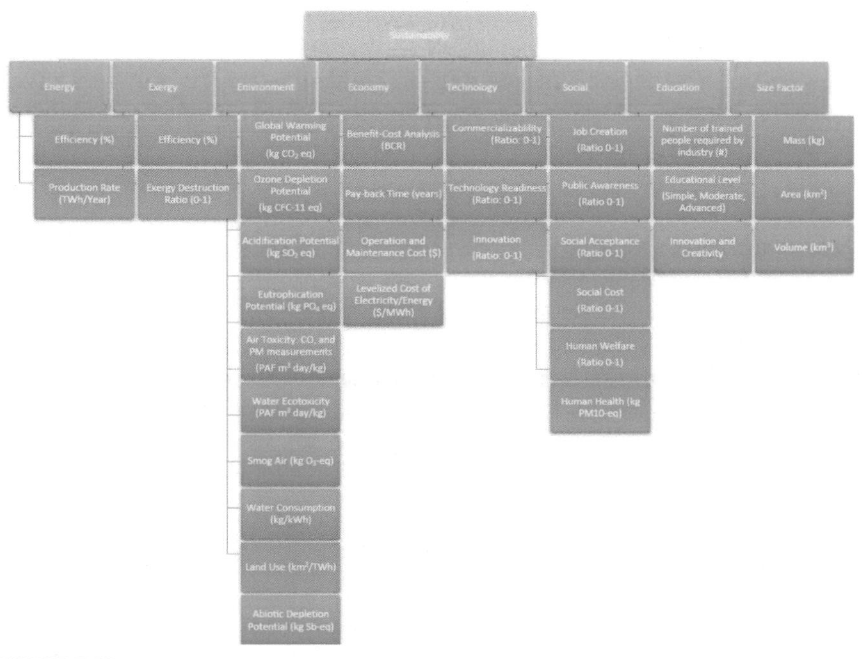

FIGURE 7.33

Integrated sustainability assessment model for energy systems.

friendliness impact and the economic impact. The energy index is assessed using the following formula:

$$Y_{ER} = (\eta \times W_{\eta}) + (Y_{Pr} \times W_{Pr}) \tag{7.19}$$

where Y_{ER} refers to the total score of this index that is calculated by the addition of the scores of the two indicators. η refers to the score of the efficiency of the energy system; W_{η} refers the weight that is given for this indicator. Y_{Pr} represents the score of the productivity of the energy system, and W_{Pr} represents the weight associated with that indicator. Furthermore, the exergy index is assessed using two main indicators: efficiency and exergy destruction. The score of this index is calculated as follows:

$$Y_{EX} = (\psi \times W_{\psi}) + (Y_{ED} \times W_{ED}) \tag{7.20}$$

where Y_{EX} represents the total score for the exergy index. The score is a function of these two indicators. ψ represents the exergy efficiency of the system, and W_{ψ} represents the allocated weight for this indicator. Y_{ED} is the score of the exergy destruction indicator and W_{ED} is the weight associated with it. In addition, the environmental friendliness index uses numerous indicators such as potential nonair environmental impacts, land use, water consumption, water quality of discharge, solid waste and ground contamination, and biodiversity. The score of this index is calculated as follows:

$$\begin{aligned} Y_{ENV} = {} & (Y_{GWP} \times W_{GWP}) + (Y_{ODP} \times W_{ODP}) + (Y_{AP} \times W_{AP}) + (Y_{EP} \times W_{EP}) \\ & + (Y_{AT} \times W_{AT}) + (Y_{WE} \times W_{WE}) + (Y_{SA} \times W_{SA}) + (Y_{WC} \times W_{WC}) \\ & + (Y_{LU} \times W_{LU}) + (Y_{ADP} \times W_{ADP}) \end{aligned} \tag{7.21}$$

where Y refers to the score for the indicators used while W refers to the weights assigned for the indicator. GWP refers to the global warming potential, ODP to the stratospheric ozone depletion potential, AP to the acidification potential, EP to the eutrophication potential, AT to air toxicity, WE to water ecotoxicity, SA to smog air, WC to water consumption, LU to the land use, and ADP to the abiotic depletion potential. These 10 indicators are carefully selected to account for all of the emissions and environmental impression that energy systems leave throughout manufacturing and operation of these systems. Further explanation follows for each indicator.

Moreover, the economic index is also taken into consideration when assessing the sustainability of this energy system. Energy systems are economically sustainable if they are profitable, serviced at lower cost for the consumer, and contain the elements of a successful business idea. The score of this index is calculated as follows:

$$Y_{ECO} = (Y_{BCR} \times W_{BCR}) + (Y_{PBT} \times W_{PBT}) + (Y_{LCOE} \times W_{LCOE}) \tag{7.22}$$

where Y_{BCR}, Y_{PBT}, and Y_{LCOE} refer to the scores of benefit-cost ratio, payback time, and the levelized cost of energy/electricity, respectively. *W* terms refer to the weight associated with each indicator. The technology index is also taken into consideration. The score of this index is calculated as follows:

$$Y_{TECH} = (Y_{COMM} \times W_{COMM}) + (Y_{TR} \times W_{TR}) + (Y_{IN} \times W_{IN}) \tag{7.23}$$

where Y_{COMM}, Y_{TR} and Y_{IN} refer to the scores of commercializability, technology readiness, and innovation. Moreover, the social index account for the social aspects associated with the energy system. The score of this index is calculated as follows:

$$Y_{SOC} = (Y_{JC} \times W_{JC}) + (Y_{PA} \times W_{PA}) + (Y_{SA} \times W_{SA}) + (Y_{SC} \times W_{SC}) + (Y_{HW} \times W_{HW}) + (Y_{HH} \times W_{HH}) \tag{7.24}$$

where Y_{JC}, Y_{PA}, Y_{SA}, Y_{SC}, Y_{HW}, and Y_{HH} refer to the scores of job creation, public awareness, social acceptance, social cost, human welfare, and human health respectively. *W* refers to the weight associated with each indicator. The educational index reflects the level of education, training, best practices, and innovation in educational methods adopted in various systems. Therefore, this index is calculated by assessing three main indicators. The score of this index is calculated as follows:

$$Y_{EDU} = (Y_{TRAIN} \times W_{TRAIN}) + (Y_{EL} \times W_{EL}) + (Y_{EI} \times W_{EI}) \tag{7.25}$$

where Y_{TRAIN}, Y_{EL}, and Y_{EI} refer to the score for the number of trained people required by the industry, educational level, and educational innovation. *W* refers to the weights associated with each indicator. Lastly, the sizing index refers to the size of the system with respect to volume, mass, and land use of the energy system. The volume and mass parameters are associated with mobile applications such as vehicles. The land use is associated with stationary applications such as a power plant. The sizing index reflects the magnitude of the energy system and consequently gives an accurate understanding when assessing its sustainability. The sizing index of the energy system in this sustainability assessment model will look at three main indicators: mass, land use, and volume. The score of this category is calculated as follows (Hacatoglu, 2014):

$$Y_{MF} = (Y_M \times W_M) + (Y_{LU} \times W_{LU}) + (Y_V \times W_V) \tag{7.26}$$

where Y_M, Y_{LU}, and Y_V refer to the score for mass, land use, and volume, respectively. *W* refers to the weights associated with each indicator.

7.2.4 Results and discussion

Thermodynamic assessment on this system is performed to understand its technical parameters as well as its energetic and exergetic performance as part of the sustainability assessment. Table 7.5 presents the main state points of the geothermal system and the thermodynamic parameters associated with each state point including the pressure, temperature, enthalpy, entropy, exergy, and position.

Table 7.5 Thermodynamic parameters of the geothermal system at each state point.

State	*P* (kPa)	*T* (K)	*h* (kJ/kg)	*s* (kJ/kg K)	Ex
0	101.3	298.2	2192	8.122	–
1	200	−18.87	1439	7.887	−87.74
2	1200	128.7	1749	7.007	187.7
3	1200	30.93	1487	7.273	149.7
4	1170	29.7	339.2	1.48	127.2
7	200	−18.87	339.2	1.771	103
7	200	−18.87	1439	7.887	−87.74

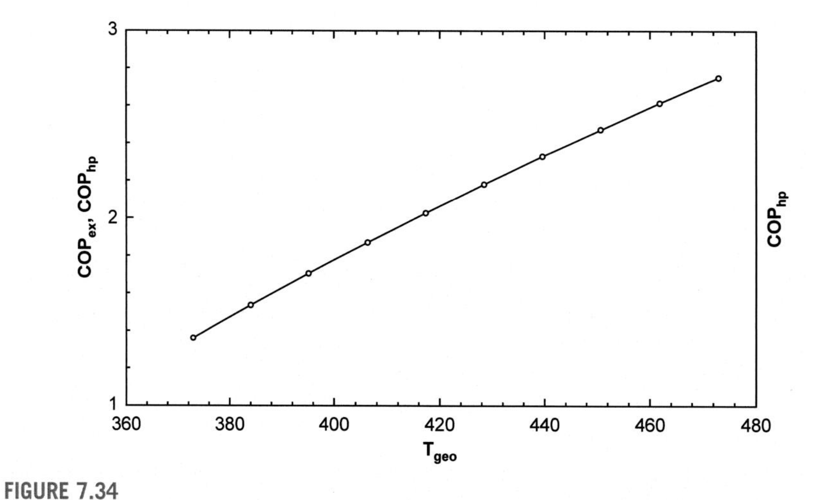

FIGURE 7.34

The impact of the geothermal fluid temperature on the exergetic and energetic COPs.

Furthermore, the influence of the geothermal fluid temperature over the exergetic COP is evident with a positive linear relationship illustrating that higher temperatures of geothermal fluid yield higher exergetic COPs. However, the COP of the heat pump remains unchanged with varying geothermal fluid temperatures. Fig. 7.34 demonstrates this relationship clearly.

Moreover, the effectiveness of the design for the geothermal heat pump system was dependent on the selection of a refrigerant with favorable thermodynamic and transport properties. Historically, the use of refrigerants has been connected to disastrous environmental consequences; thus, significant consideration was also given to the global warming potential of the refrigerant (Fig. 7.35).

Analysis of a vapor compression refrigeration cycle with R134a, R1234yf, and ammonia revealed ammonia to be the most favorable refrigerant in terms of its

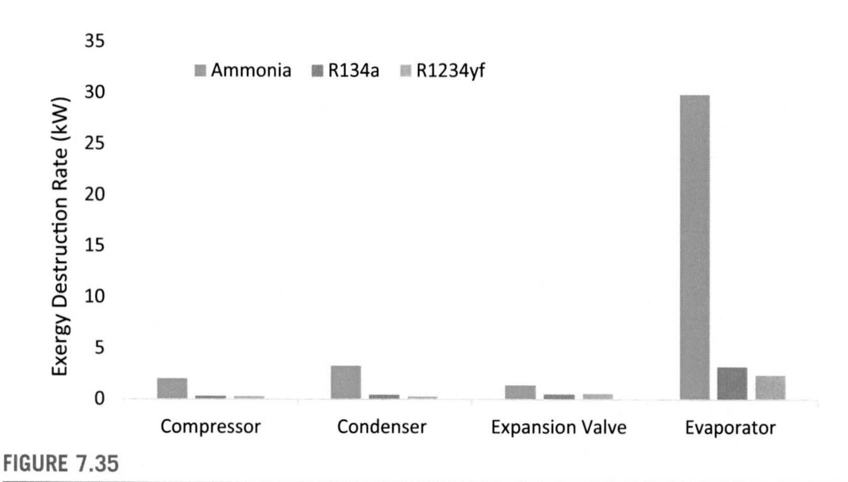

FIGURE 7.35

Exergy destruction rates of the main components of the heat pump system using different refrigerants.

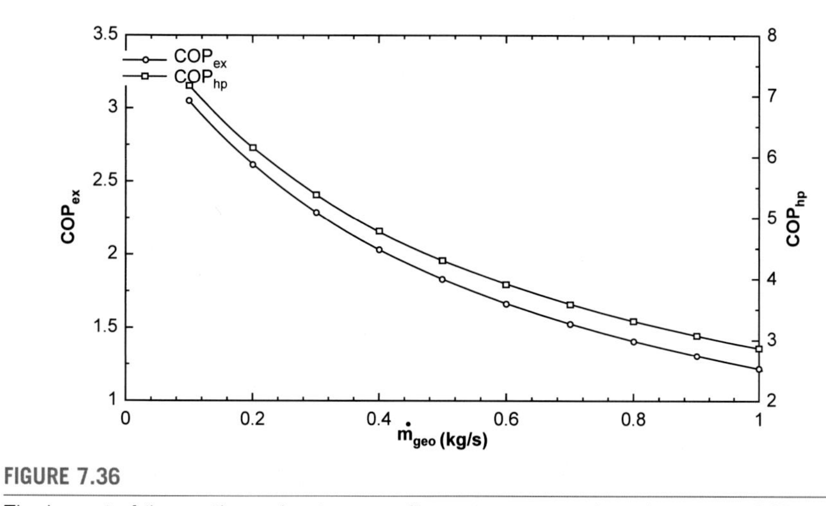

FIGURE 7.36

The impact of the geothermal water mass flow rate on exergetic and energetic COPs.

thermodynamic and transport properties. R134a was eliminated as it is currently being phased out due to its 100 year GWP. R1234yf is the least effective in terms of its thermodynamic and transport properties (Fig. 7.36).

This investigation and comparison of different refrigerants revealed the considerable potential and current need for the continued development of refrigerant, which are both environmentally benign and thermodynamically effective. Fig. 7.37 shows the exergy destruction of each refrigerant at each stage of the heat pump cycle.

The evaporator hosts the most exergy destruction throughout the different refrigerants with ammonia having a significant exergy destruction at this stage. On a different note, the impact of the mass flow rate of the geothermal fluid on exergetic and energetic COPs is presented in Fig. 7.37. The trend shows a negative exponential curve: as the mass flow rate increases, both the energetic and exergetic COPs decrease in a similar fashion. Moreover, the exergetic and energetic COPs are assessed for each refrigerant. Fig. 7.37 shows the outcome, which shows the highest exergetic COP to be that of R1234yf. In addition, the *T*–*s* diagram is presented in Fig. 7.38 showing the different state points and their position with respect to temperature and entropy.

As for the solar system, the ambient temperature was analyzed for its impact on the exergy destruction of the solar system as well as the exergy efficiency. It is

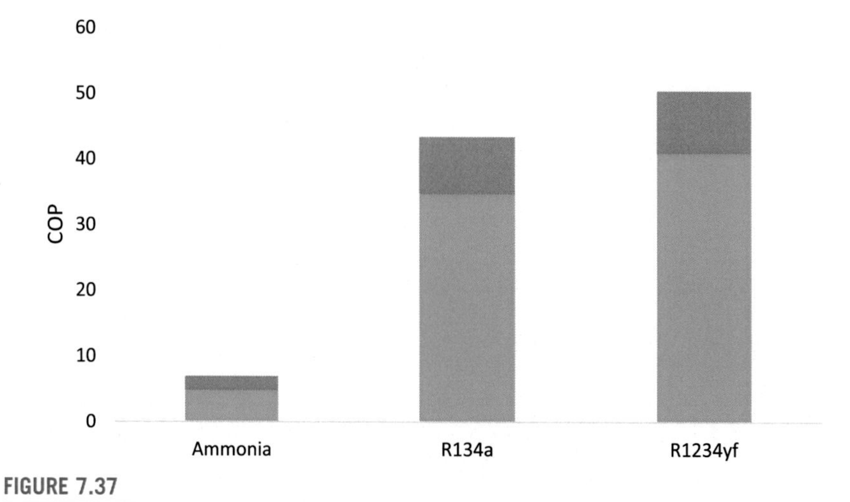

FIGURE 7.37

Exergetic (orange) and energetic (blue) COPs using different refrigerants.

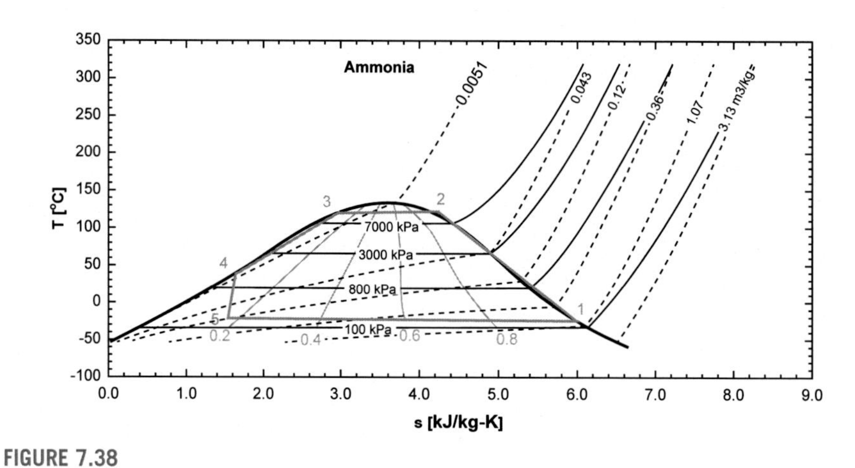

FIGURE 7.38

T–*s* diagram of the heat pump system using ammonia refrigerant.

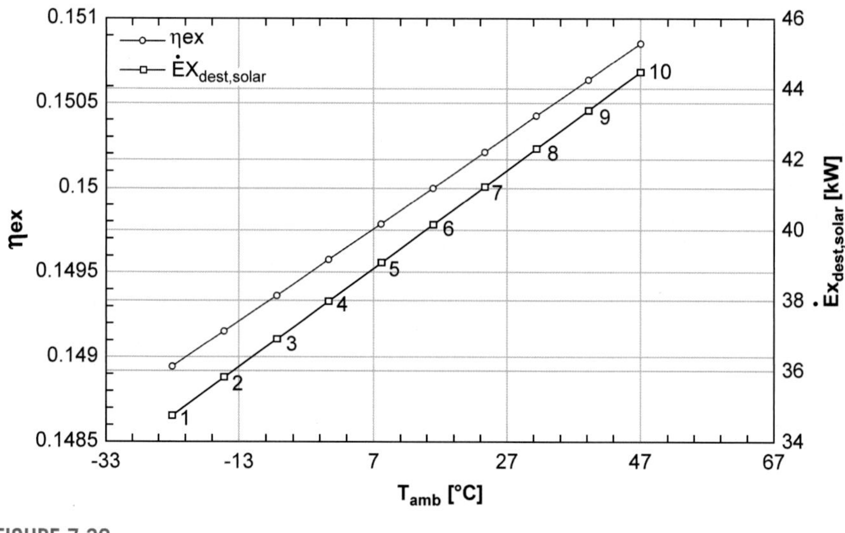

FIGURE 7.39

Impact of ambient temperature on the exergy destruction of the solar system and the exergy efficiency.

evident from Fig. 7.39 that the relationship between these two variables and the ambient temperature is that of a positive linear relationship. As the ambient temperature increases, the exergy efficiency of the system also increases linearly; however, the exergy destruction also increases in a similar fashion.

On the other hand, the irradiance has a different trend on the exergy destruction and the solar heat output. As shown in Fig. 7.40, as the solar irradiance increases, the exergy destruction increases in a linear trend. However, the solar heat output also increases in a linear, yet steeper trend than that of the exergy destruction.

7.2.4.1 Sustainability assessment results

The time-space-receptor method was also used in appointing appropriate values for each index. The indexes used in this study vary as some have long-term impact such as exergy and energy indexes while others have short-term impact. Table 7.6 shows the various indexes and their associated weights as per the schemes used: panel method, individualist, egalitarian, hierarchist, and equal weighting method.

There are slight variations between the different schemes in prioritizing specific indexes over other indexes as illustrated in Fig. 7.41. For example, the panel method prioritized the exergy index and neglected the sizing index while the individualist method prioritized the social index and neglected technology index.

Furthermore, the panel gave less priority for education whereas all the other schemes placed it at a higher priority compared to the panel scheme. When analyzing the overall system, the sustainability index ranges between 0.77 and

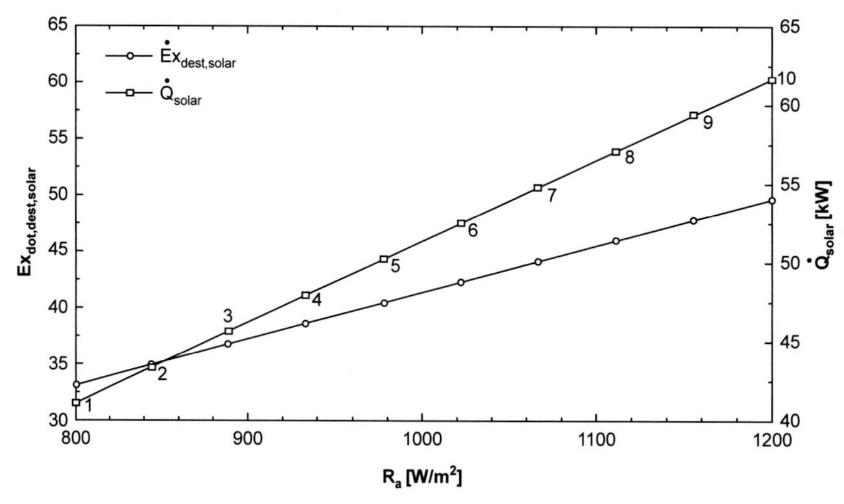

FIGURE 7.40

Impact of solar irradiance on the exergy destruction of the solar system and the solar energy output.

Table 7.6 Priority factors of the eight main indexes of the sustainability model with respect to various schemes.

Index	Individualist	Egalitarian	Hierarchist	Panel	Equal weighting
Energy	0.13	0.12	0.13	0.10	0.13
Exergy	0.13	0.12	0.13	0.17	0.13
Environmental friendliness	0.13	0.13	0.17	0.18	0.13
Economic	0.13	0.17	0.13	0.14	0.13
Technology	0.09	0.12	0.12	0.12	0.13
Social	0.17	0.12	0.12	0.17	0.13
Educational	0.13	0.12	0.12	0.09	0.13
Sizing	0.10	0.10	0.12	0.07	0.13

0.77 depending on which aggregation scheme is used. Fig. 7.42 illustrates the sustainability assessment results of the system with the value 1 being the highest value for sustainability.

It is evident that the equal weighting aggregation method yields higher sustainability values while the individualist aggregation method yields lower sustainability values. The purpose behind using different aggregation methods is to reach to an accurate result, which can be considered reliable. It is also used to understand the

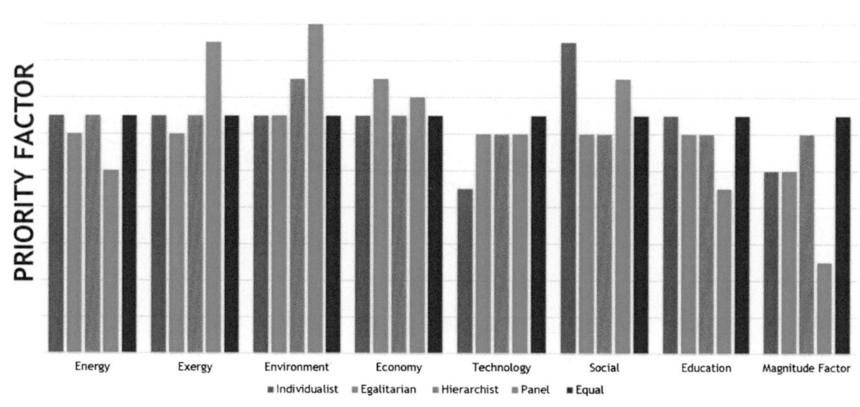

FIGURE 7.41

Distribution of priority factors based on the four schemes for the main indexes used for the sustainability assessment model.

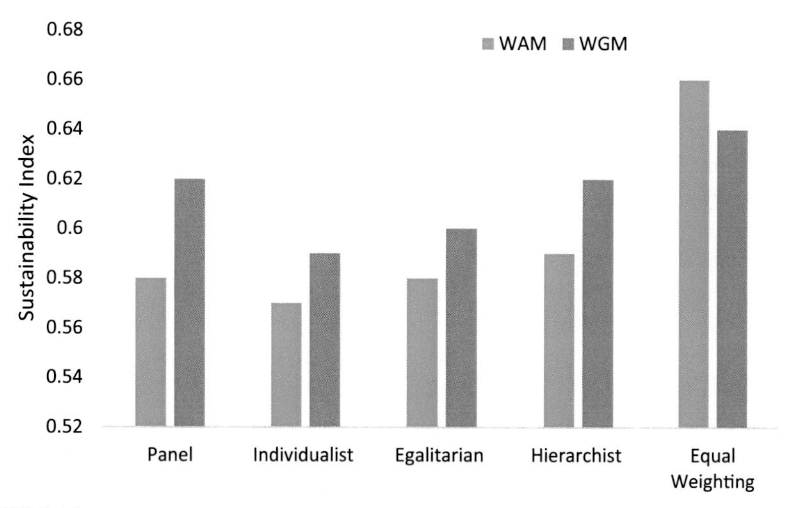

FIGURE 7.42

Distribution of the sustainability index results based on the various aggregation method and characterization scheme.

degree of variance between the results. In this case, the results vary with a maximum of 0.09 points, which if converted into percentage; it is a 9% variance.

The egalitarian scheme seems to be most moderate, and thus to analyze it further, the top three indexes have been investigated in depth. Fig. 7.43 shows the impact of each of these indexes, as their value increases from 0.2 to 1 on the overall sustainability index. It is evident that manipulating the economic index yields the highest

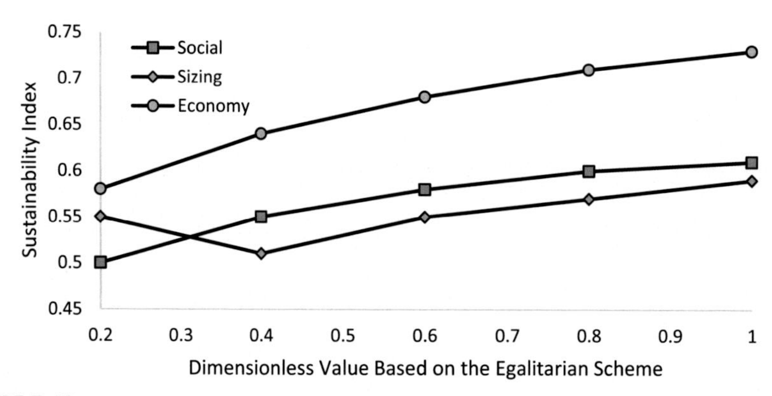

FIGURE 7.43

Comparison of the main indexes using the egalitarian scheme on the overall sustainability index.

change over the sustainability index. However, the sizing index does not influence the overall sustainability index as much.

7.2.5 Closing remarks

In conclusion, the sustainability assessment of a net zero energy house is conducted. The analysis of the two renewable energy sources included balanced equations for mass, energy, and exergy for both systems and their subcomponents. Energetic and exergetic COPs are calculated. EES is used to perform parametric studies to vary the operating and environmental conditions to determine the effect on system performance. The geothermal heat pump system is designed to satisfy a heating load of 7.7 kW using ammonia as the refrigerant. The COP of the geothermal heat pump is calculated to be 4.9 while the exergetic COP is 2.1. The effect of the condenser pressure on the performance of the heat pump is also analyzed. Increasing the condenser pressure increased the temperature of the refrigerant, leaving the condenser; however, higher pressures required higher compressor work input and decreased the heat output. Overall, increasing the condenser pressure lowered the cycle COP. The performance of the heat pump system using three different refrigerants, R1234yf, R134a, and ammonia, is also investigated. In addition to their thermodynamic properties, the environmental impact of each refrigerant is considered. Ammonia is selected as the most appropriate due to its nontoxicity and low 100 GWP despite inferior COP results. Furthermore, the yielded electricity production from the solar system is 71.4 kW with exergetic efficiency of 17% under atmospheric pressure and temperature. The sustainability index of the system ranges between 77% and 77%. The manipulation of the economic index using the egalitarian aggregation method could yield a sustainability index of up to 73%.

7.3 Case study 3: sustainability assessment of an integrated solar PVT system

The objective of this case study is to provide efficient, clean, and dependable energy options to meet the residential demand for electricity, heating, cooling, and hot water for 170 Ontarian homes. The reason behind this number is that it approximates the number of homes in a village or town. Solar photovoltaic technology is used in this study.

7.3.1 Introduction

To meet the demand, the total energy production is 7.13 GWh. The case study is based on developments based in Ontario, Canada with 70% of south-facing roofs, 27% east-facing roofs, and 27% west-facing roofs. The total size of the system is calculated using the following equation:

$$\dot{S}S = \frac{\text{PR}}{\text{YP}_{\text{residential}}} \tag{7.27}$$

where PR is the production rate and $\text{YP}_{\text{residential}}$ is the yield production for residential purposes, which is set to be 1200 kWh/kWp as per the industrial standards. Therefore, the total size of the system is 4.38 MW with approximately 13,781 solar panels to be installed. This comes to an average of 91 panels per household with a power rate of 0.32 kW/panel. However, it is anticipated that this system would be installed on roofs of spacious warehouses and commercial buildings. The area needed to install this system is approximately 7 acres, which keeps in consideration the shading effect. The area for the system is calculated using the following equation:

$$\eta_{\max} = \frac{P_{\max}}{E_{S,Y}^{SW} \times A_c} \tag{7.28}$$

where $\eta_{\max}$ is the maximum efficiency, which is set at 20%, $P_{\max}$ is the maximum power output for the target energy demand, which is 1.27 MW, $E_{S,Y}^{SW}$ is the incident radiation flux, which is set at 1 kW/m^2 for the purposes of this study, and A_c is the area of collector in square meter. Fig. 7.44 shows the general layout of the solar-PV system with a heat pump. The type of silicon used to produce the solar cells has an impact on the power conversion efficiency of a PV module. As observed in Fig. 7.44, the solar radiation received by the PV module is converted into electricity from the initial state of phonic energy.

This way, the PV modules produce DC power, which is connected to a DC motor that is also linked with a compressor. Matching the electrical properties of the motor and the available voltage and current generated by the PV modules becomes the main concern.

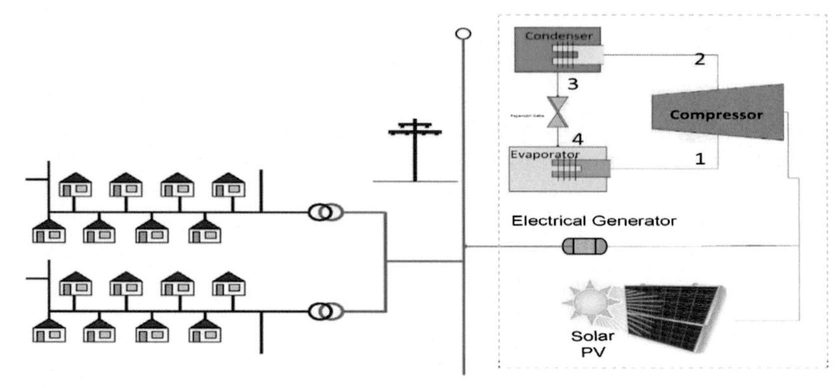

FIGURE 7.44

A schematic sketch illustrating the analyzed solar PV case study.

7.3.2 System analysis

EES is used, and a model is developed for annual simulation of this system with input variables described in Table 7.7. Indeed, solar irradiance is also accounted in this system analysis.

Table 7.7 Input parameters used to assess the photovoltaic system used in this case study.

Parameter	Value	Reference
Liveable floor space	197 m^2	Saldanha and Beausoleil-Morrison (2012)
Coefficient of performance of an average central air conditioning system	4	Sandler (1999)
Effectiveness of regenerator	0.77	Cengel et al. (2010)
Efficiency of combustion	0.87	Sandler (1999)
Electric generator efficiency	0.92	Zini and Tartarini (2010)
Isentropic efficiency of a compressor	0.77	Cengel et al. (2010)
Isentropic efficiency of a gas turbine	0.77	Cengel et al. (2010)
Isentropic efficiency of a pump	0.77	Cengel et al. (2010)
Number of people per household	4	Saldanha and Beausoleil-Morrison (2012)
Space-heating factor	0.7 W/m^2 K	Sørensen (2011)
Temperature of domestic hot water	70°C	Sandler (1999)
Temperature of household	18°C	Saldanha and Beausoleil-Morrison (2012)

Source: Reproduced from Hacatoglu, 2016.

7.3.3 Solar irradiance data

According to Natural Resources Canada (2007), the solar irradiance in Ontario is promising. Solar energy potential is measured in kWh generated per kW of the installed photovoltaic capacity. South-facing rooftops can ultimately harvest solar energy to meet the residential demand. Ontario municipalities vary greatly in their irradiance rates. For example, the annual PV potential for Toronto is 1171 kWh/kW, which is decent when compared to Cairo, Egypt at 1737 kWh/kW.

7.3.4 Solar modeling

The sequence of these equations follows the schematic sketch presented in Fig. 7.44. The compressor is labeled as state number 1 followed by the condenser, the valve, and finally the evaporator as state number 4.

For the adiabatic compressor, one can write the thermodynamic balance equations as follows:

$$\text{Mass Balance Equation (MBE):} \quad \dot{m}_1 = \dot{m}_2 \tag{7.29}$$

$$\text{Energy Balance Equation (EBE):} \quad \dot{m}_1 h_1 + \dot{W}_{in} = \dot{m}_2 h_2 \tag{7.30}$$

$$\text{Entropy Balance Equation (EnBE):} \quad \dot{m}_1 s_1 + \dot{S}_{gen} = \dot{m}_2 s_2 \tag{7.31}$$

$$\text{Exergy Balance Equation (ExBE):} \quad \dot{m}_1 ex_1 + \dot{W}_{in} = \dot{m}_2 ex_2 + \dot{E}x_{dest} \tag{7.32}$$

For the condenser, one can write the thermodynamic balance equations as follows:

$$\text{Mass Balance Equation (MBE):} \quad \dot{m}_2 = \dot{m}_3 \tag{7.33}$$

$$\text{Energy Balance Equation (EBE):} \quad \dot{m}_2 h_2 = \dot{m}_3 h_3 + \dot{Q}_{out} \tag{7.34}$$

$$\text{Entropy Balance Equation (EnBE):} \quad \dot{m}_2 s_2 + \dot{S}_{gen} = \dot{m}_3 s_3 + \frac{\dot{Q}_{out}}{T_0} \tag{7.35}$$

$$\text{Exergy Balance Equation (ExBE):} \quad \dot{m}_2 ex_2 = \dot{m}_3 ex_3 + \dot{Q}_{out}\left(1 - \frac{T_0}{T_0}\right) + \dot{E}x_{dest} \tag{7.36}$$

For the evaporator, one can write the thermodynamic balance equations as follows:

$$\text{Mass Balance Equation (MBE):} \quad \dot{m}_4 = \dot{m}_1 \tag{7.37}$$

$$\text{Energy Balance Equation (EBE):} \quad \dot{m}_4 h_4 + \dot{Q}_{in} = \dot{m}_1 h_1 \tag{7.38}$$

$$\text{Entropy Balance Equation (EnBE):} \quad \dot{m}_4 s_4 + \frac{\dot{Q}_{in}}{T_{space}} + \dot{S}_{gen} = \dot{m}_1 s_1 \tag{7.39}$$

$$\text{Exergy Balance Equation (ExBE):} \quad \dot{m}_4 ex_4 + \dot{Q}_{in}\left(1 - \frac{T_0}{T_{space}}\right) = \dot{m}_1 ex_1 + \dot{E}x_{dest} \tag{7.40}$$

For the expansion valve, one can write the thermodynamic balance equations as follows:

$$\text{Mass Balance Equation (MBE):} \quad \dot{m}_3 = \dot{m}_4 \tag{7.41}$$

$$\text{Energy Balance Equation (EBE):} \quad \dot{m}_3 h_3 = \dot{m}_4 h_4 \tag{7.42}$$

$$\text{Entropy Balance Equation (EnBE):} \quad \dot{m}_3 s_3 + \dot{S}_{gen} = \dot{m}_4 s_4 \tag{7.43}$$

$$\text{Exergy Balance Equation (ExBE):} \quad \dot{m}_3 ex_3 = \dot{m}_4 ex_4 + \dot{E}x_{dest} \tag{7.44}$$

where the process is isenthalpic and entropy at state 4 can be found using enthalpy and pressure parameters.

The PV system analysis is modeled using the following equations:

$$\dot{Q}_{Solar}\text{Area}_{PV} = \dot{W}_{PV} + \dot{Q}_{Loss,PV} \tag{7.45}$$

$$n_{\ PV} = \frac{\dot{W}_{PV}}{\dot{Q}_{Solar}\text{Area}_{PV}} \tag{7.46}$$

$$\frac{\dot{Q}_{Loss,PV}}{T_{amb}} = \frac{\dot{Q}_{Solar}}{T_{Sun}}\text{Area}_{PV} + \dot{S}_{G,PV} \tag{7.47}$$

$$\dot{E}x_{Q,Solar} = \dot{W}_{PV} + \dot{E}x_{Q,Loss,PV} + \dot{E}x_{D,PV} \tag{7.48}$$

$$\dot{E}x_{Q,Loss,PV} = \dot{Q}_{Loss,PV}\left(1 - \frac{T_0}{T_{amb}}\right) \tag{7.49}$$

$$\dot{E}x_{Q,Solar} = \dot{Q}_{Solar}\text{Area}_{PV}\left[1 - \frac{4}{3}\left(\frac{T_0}{T_{amb}}\right) + \frac{1}{3}\left(\frac{T_0}{T_{amb}}\right)^4\right] \tag{7.50}$$

7.3.5 Results and discussion

The time-space-receptor method was also used in appointing appropriate values for each index. The indexes used in this study vary as some have long-term impact such as exergy and energy indexes while others have short-term impact. Table 7.8 shows the various indexes and their associated weights as per the schemes used: panel method, individualist, egalitarian, hierarchist, and equal weighting method.

There are slight variations between the different schemes in prioritizing specific indexes over other indexes as illustrated in Fig. 7.41. For example, the panel method prioritized the exergy index and neglected the sizing index while the individualist method prioritized the social index and neglected technology index. Furthermore,

Table 7.8 Priority factors of the eight main indexes of the sustainability model with respect to various schemes.

Index	Individualist	Egalitarian	Hierarchist	Panel	Equal weighting
Energy	0.13	0.12	0.13	0.10	0.13
Exergy	0.13	0.12	0.13	0.17	0.13
Environmental friendliness	0.13	0.13	0.17	0.18	0.13
Economic	0.13	0.17	0.13	0.14	0.13
Technology	0.09	0.12	0.12	0.12	0.13
Social	0.17	0.12	0.12	0.17	0.13
Educational	0.13	0.12	0.12	0.09	0.13
Sizing	0.10	0.10	0.12	0.07	0.13

Table 7.9 Final sustainability index of PV system using various weighting and aggregation methods.

Scheme	Sustainability index based on WGM	Sustainability index based on WAM
Panel	0.77	0.79
Individualist	0.77	0.74
Egalitarian	0.78	0.77
Hierarchist	0.79	0.77
Equal weighting	0.79	0.73

the panel gave less priority for education whereas all the other schemes placed it at a higher priority compared to the panel scheme.

Because weighting and aggregation are subjective concepts, various weighting schemes and aggregation methods have been used to minimize any noise and to present results objectively as much as possible. Therefore, there are two aggregation schemes used, which are weighting arithmetic mean and weighting geometric mean. Table 7.9 shows the final sustainability index for the PV system with respect to the characterization schemes applied and the aggregation method used.

From Table 7.9, it evident that there is a slight difference between the uses of weighted geometric mean (WGM) and the weighted arithmetic mean (WAM) for calculating the sustainability index of the PV system. Using WAM, the sustainability index derived from the equal weighting and panel schemes was higher than the derived values using WGM. Fig. 7.45 shows the plot, which compares all values from various methods. The difference of values between the WGM and the WAM for the hierarchist and equal weighting schemes double the difference in values

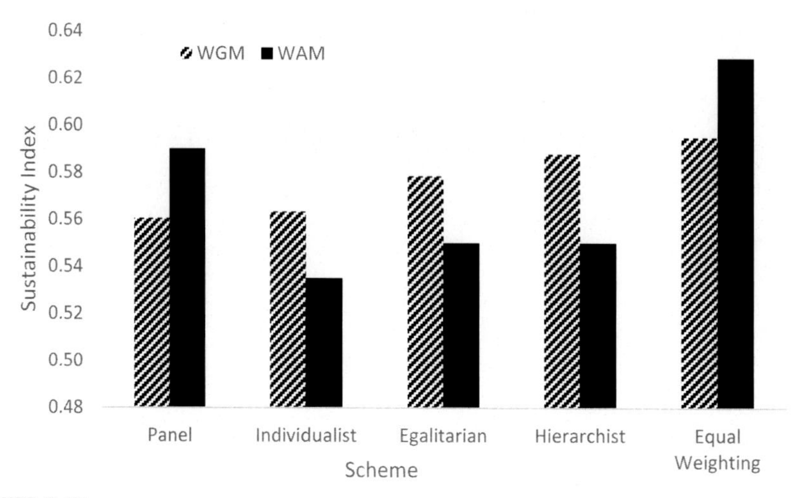

FIGURE 7.45

Distribution of the sustainability index results based on the various aggregation method and characterization scheme for the solar PV energy system.

for the panel, individualist and egalitarian schemes. On the contrary, it is apparent that the sustainability index derived from the other weighting schemes when using WAM was less than the values obtained when using WGM. Moreover, the graph illustrates the weak points of the WAM and highlights the advantage of using the WGM over the WAM when assessing sustainability of PV systems. Moreover, the final sustainability index using the WGM varied by a factor of 0.03 between the different weighting schemes as illustrated in Fig. 7.46. Furthermore, the correlated difference between the lowest value with respect to the highest is 0.97.

The individualist and panel methods yielded a similar score while equal weighting suggested the highest sustainability score for the PV system in this case study. The minimum variation that exists in these results suggests its robustness and strength over the results extracted using the WAM as observed in Fig. 7.47.

In this illustration, the panel method is similar to the equal weighting method, which differs from WGM where the panel method was parallel to the individualist method. On the other hand, the individualist, egalitarian, and hierarchist schemes resulted in very similar sustainability index values, which are relatively lower than the other two schemes. Furthermore, the variation in data in the WAM is triple times the variation in the WGM. This suggests that the WAM is less preferable when aggregating values for the sustainability assessment. On another note, it is critical to know how changes in the inputs affect the final sustainability index.

As observed earlier in Table 7.9, each characterization scheme highlights a number of selected categories. For example, using the individualist scheme, the technology and social categories are the first two categories that catch attention. This is because the social category received the highest priority factor while the technology category received the lowest. Therefore, using this methodology, the following

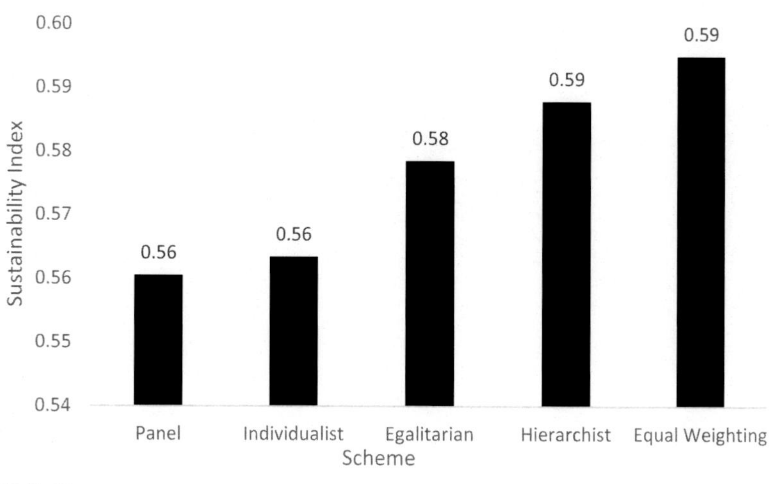

FIGURE 7.46

Distribution of the sustainability index results based on WGM and characterization schemes for the solar PV system.

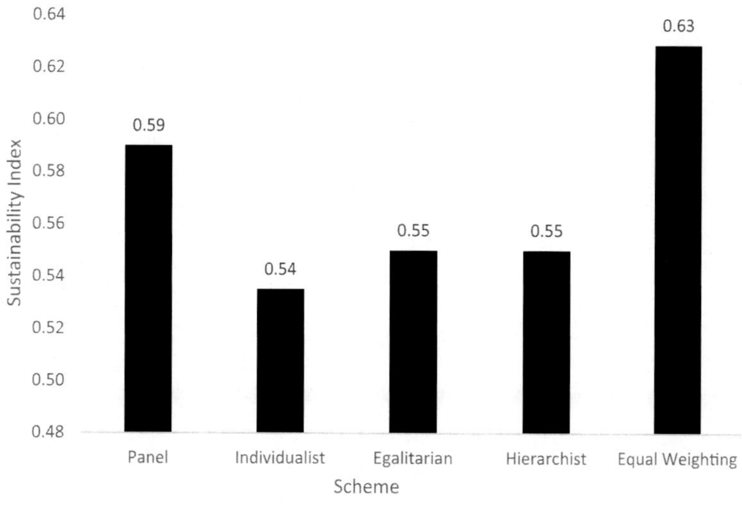

FIGURE 7.47

Distribution of the sustainability index results based on WAM and characterization schemes for the solar PV system.

figures will help us analyze the influence of these selected categories on the final sustainability index of the solar PV system. When conducting this, values of all other categories were kept as is to account only for the effect of the category of interest on the sustainability index score. Furthermore, the results were extracted using

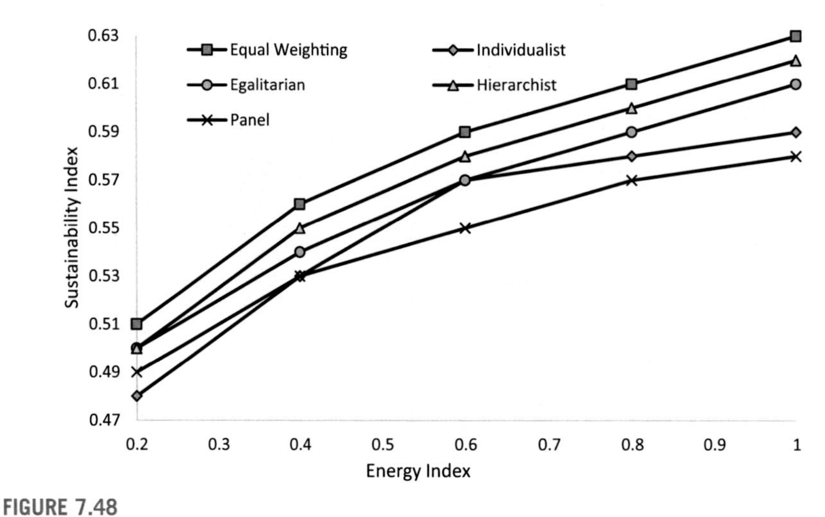

FIGURE 7.48

Variation of the solar PV sustainability index with respect to the energy index based on WGM.

the WGM. Fig. 7.48 shows the various schemes and the influence of the energy category as its score increases from 0.2 to 1 (1 being the most favorable).

Equal weighting and hierarchist schemes show a similar trend for the energy index, characterized by steady increase of the sustainability index as the energy index increases. This also goes with the egalitarian scheme. However, the panel method and the individualist scheme present slightly different dynamics. As for the individualist scheme, the energy index is most impactful on the sustainability index when the value is between 0.2 and 0.7, after which the increase results in a less steeper and more steadily increase in the sustainability index value. Similarly, the panel method shows that the energy index is most impactful on the sustainability index when the value increases from 0.2 to 0.4. The most impactful index refers to the index that if increases in value, the overall system achieves higher sustainability index. In specific, the changes of the energy index values have a higher degree of influence on the overall sustainability index scores as opposed to other indexes. Then, as the value of the energy index increases, its effect is steady.

It is critical to mention that if the value of any index is zero, that will automatically result in a sustainability index of zero as well. This is because the weighted geometric mean is the product and not the sum of all of these indices. Therefore, if any index has a value of zero, the sustainability index consequently results in a value of zero. Fig. 7.49 presents the exergy index and its impact on the sustainability index. Similar to the energy index, the equal weighting scheme suggests the highest sustainability index scores as the exergy index increases. The sustainability index score changes between 0.77 and 0.48 depending on the exergy index performance. This accounts for 17% increase or decrease on the sustainability index, all due to the exergy index performance.

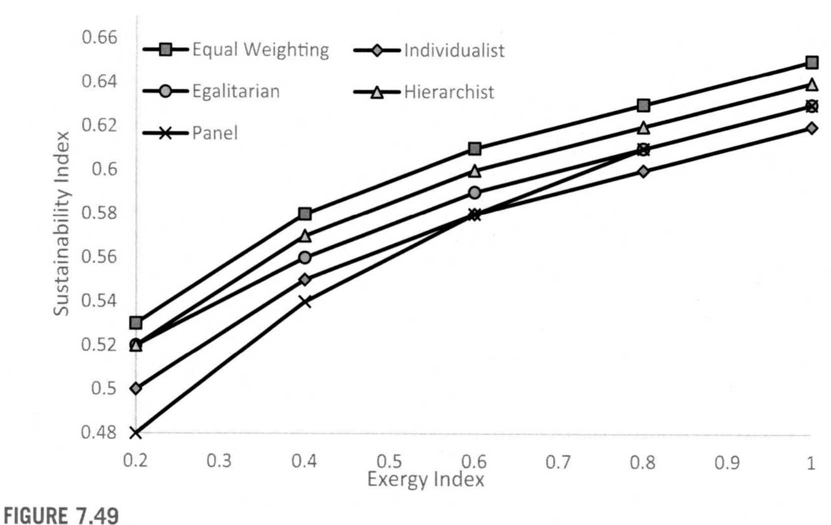

FIGURE 7.49

Variation of the solar PV sustainability index with respect to the exergy index based on WGM.

Similar dynamics on the impact of the panel and individualist method for the energy index are also observed for the exergy index. Fig. 7.50 displays the environmental friendliness index and its influence on the sustainability index.

According to the panel method, the environmental friendliness index changes between 0.2 and 0.4 and have a minimal effect. As the score move between 0.4 and 1, its impact on the sustainability index is much higher. In fact, this increase alone accounts for 10% increase on the sustainability index with scores varying from 0.72 to 0.72. Fig. 7.51 illustrates the impact of the economic index on the sustainability index.

The changes in the economic index from 0.2 to 0.4 are presented on a steeper line compared to value changes from 0.4 to 1. Although the panel method and individualist scheme show a similar relationship, the other three schemes also present a steady increase of sustainability index due to steady increase of the economic index. Fig. 7.52 illustrates the variation of sustainability index with respect to the changes in the technology index. Most schemes show that technology index changes between 0.2 and 0.4 yield the highest rate of change on the sustainability index, which translates in larger increase segment. After that, steady increase is observed except for the values resulting from the individualist scheme as the line becomes less steep as the technological index increases from 0.8 to 1. Furthermore, the equal weighting scheme followed by the hierarchist and egalitarian yield the highest sustainability index scores for the technology index's increase in value. Fig. 7.53 presents the results of the social category.

Fig. 7.53 shows that the sustainability index scores are more variable among the schemes when the social index is low. As the social index reaches toward 0.8 and 1,

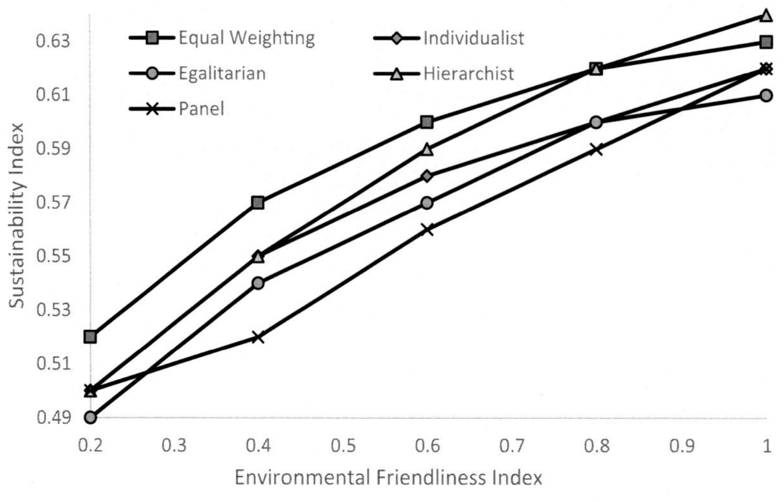

FIGURE 7.50

Variation of the solar PV sustainability index with respect to the environmental friendliness index based on WGM.

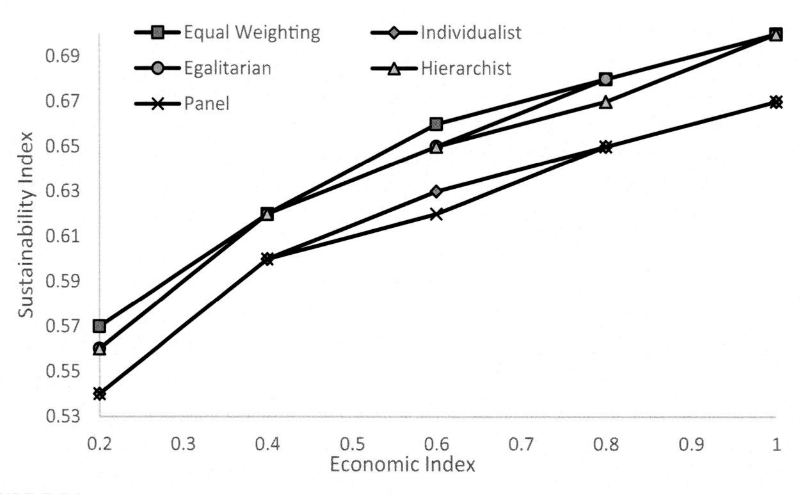

FIGURE 7.51

Variation of the solar PV sustainability index with respect to the economic index based on WGM.

variability among schemes on the sustainability index score is reduced. This means that consensus is achieved in the decision-making criteria between all schemes for that value range. Fig. 7.54 shows the educational index and its effect on the sustainability index.

The lowest sustainability index score resulting from this index's poor performance is 0.49 according to the individualist scheme while the highest score is

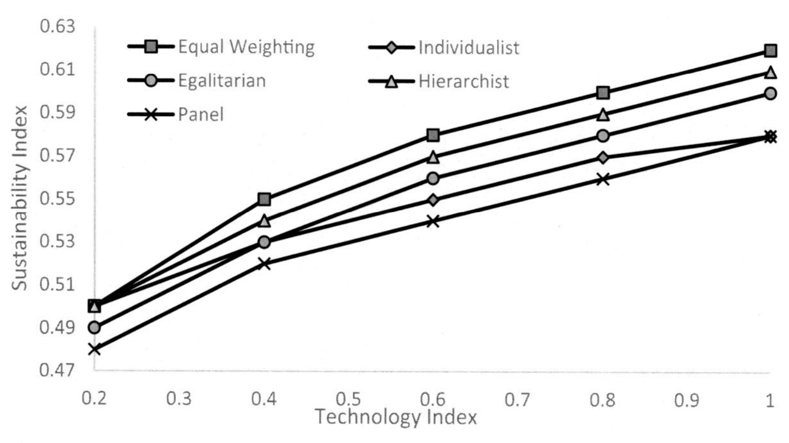

FIGURE 7.52

Variation of the solar PV sustainability index with respect to the technology index based on WGM.

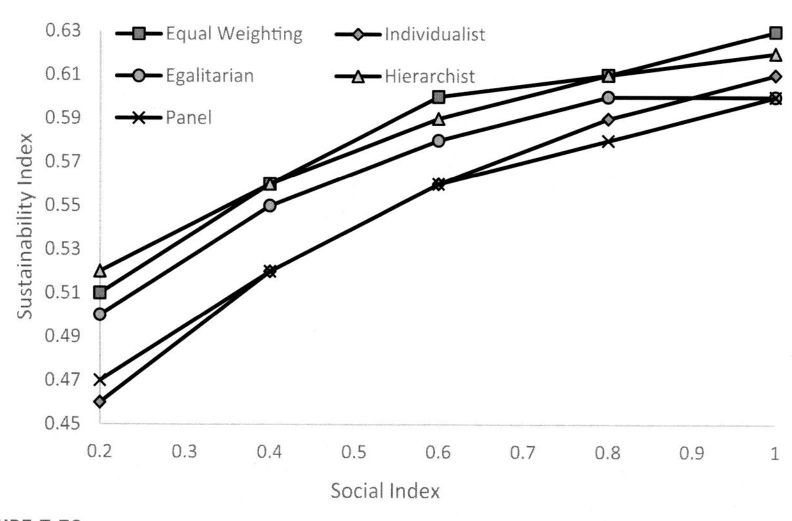

FIGURE 7.53

Variation of the solar PV sustainability index with respect to the social index based on WGM.

0.73 according to the equal weighting scheme. According to the egalitarian scheme, the sustainability index saturates after reaching 0.8 and does not increase. This is associated with the degree of influence of the social index based on this scheme. Fig. 7.55 shows the impact of the sizing index and its variation on the sustainability index.

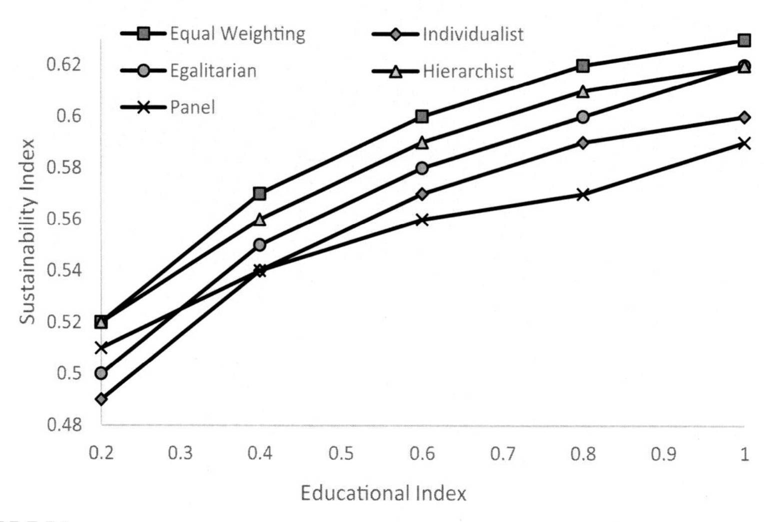

FIGURE 7.54

Variation of the solar PV sustainability index with respect to the educational index based on WGM.

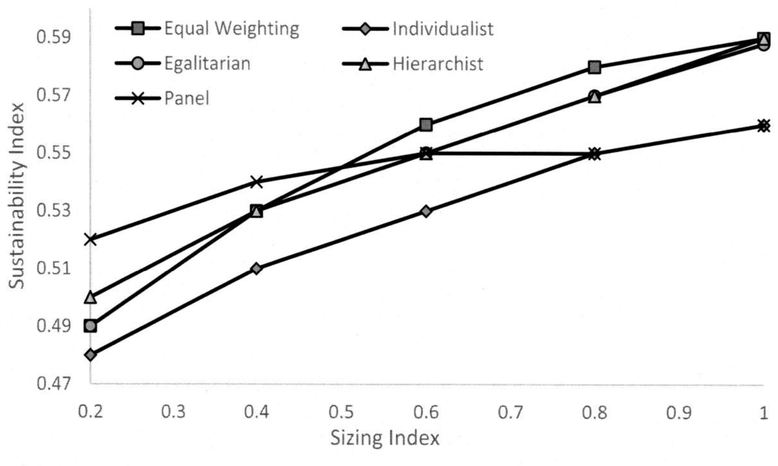

FIGURE 7.55

Variation of the solar PV sustainability index with respect to the sizing index based on WGM.

Interestingly, according to the panel method, increase of performance of the sizing index from 0.7 to 0.8 results in the same sustainability index score. Furthermore, equal weighting method results in the second lowest sustainability index score when the sizing index is low. However, when it surpasses 0.7, the equal weighting method results in the highest sustainability index score.

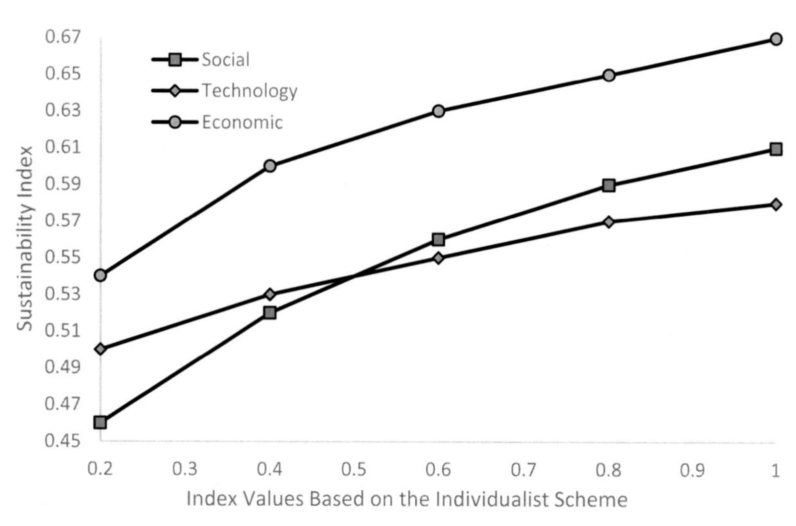

FIGURE 7.56

Variation of the solar PV sustainability index with respect to various indexes based on the individualist scheme.

According to Table 7.9, social, technology, and environment indexes were the most prominent and had unique priority factors with respect to the individualist weighting scheme. Their impact on the sustainability index is presented in Fig. 7.56.

According to the individualist scheme, the social index is the most important among all other indexes. However, it is observed that social index values lower than 0.4 result in the lowest sustainability index scores with respect to the other indexes. Social index values higher than 0.4 cause steady increase in sustainability index. However, it is evident that the economic index is the most critical when it comes to impact on the sustainability index. Therefore, although the social index was given a higher priority by the individualist scheme in previous illustrations, the economic index is in fact more important and critical to the final sustainability index. Lastly, technology index was rated the lowest according to the individualist scheme and thus its effect on the sustainability index is observed through a less steep and linear graph. Fig. 7.57 highlights key indexes related to the egalitarian scheme and the sustainability index variations.

According to the egalitarian scheme, economic index was considered the most important index followed by other indexes such as the social index while the sizing index was considered the least important one effecting sustainability of solar PV. Fig. 7.57 clearly resembles this categorization and the relationship of each index to the sustainability index score. In this figure, economic index yields the highest sustainability index scores followed by social and lastly the sizing indexes. Similar to Fig. 7.53, the social index's influence based on this scheme saturates once the sustainability index reaches 0.8. Figure 7.58 highlights the hierarchist scheme and the key indexes related to this scheme with respect to the

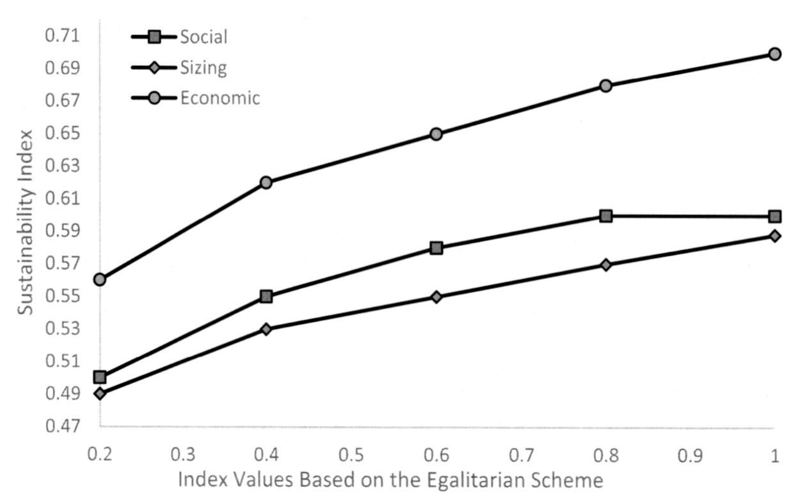

FIGURE 7.57

Variation of the solar PV sustainability index with respect to various indexes based on the egalitarian scheme.

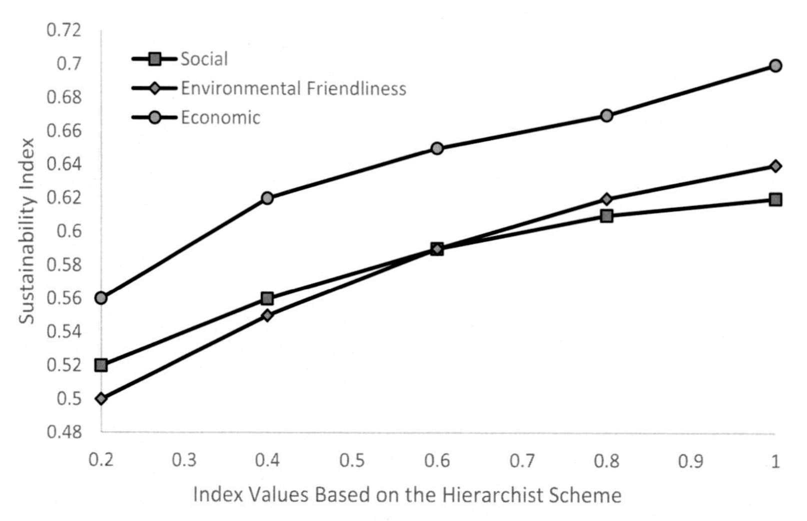

FIGURE 7.58

Variation of the solar PV sustainability index with respect to various indexes based on the hierarchist scheme.

sustainability index. The hierarchist scheme considers the environmental friendliness index to be the most important followed by economic and social indexes. In Fig. 7.58, it is obvious that the changes in economic index result in a higher sustainability index score as opposed to social or environmental friendliness indexes.

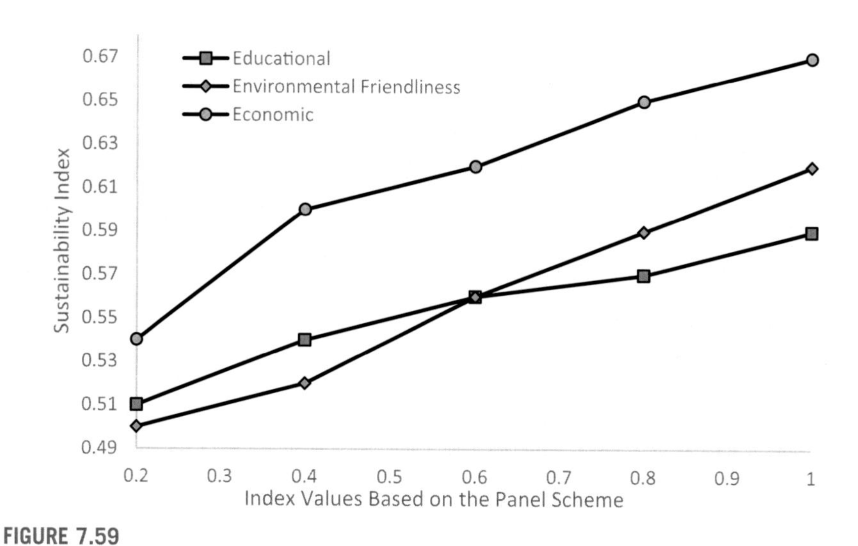

FIGURE 7.59

Variation of the solar PV sustainability index with respect to various indexes based on the panel scheme.

The economic index line is steeper, and then becomes steadier followed by another steepness as values approach 1.

Moreover, the environmental friendliness index is ranked second in yielding higher sustainability index score; however, that is only true when the environmental friendliness index value is higher than 0.4. If not, the social index becomes the second in the ranking. Fig. 7.59 presents the panel method and the variation of sustainability index scores with respect to key indexes.

The panel rated the environmental friendliness index to be the most important of all indexes with priority factor of 0.18 while the economic index was 0.14 and the educational index was 0.09. Fig. 7.57 shows conflicting results compared to the importance ranking. Although the environmental friendliness index has the highest priority, the economic index is the one that yields higher sustainability index scores with steeper effect when the value is between 0.2 and 0.4. On the other hand, the environmental friendliness index is only effective when its value is higher than 0.4; otherwise, it yields the lowest sustainability index out of these indexes presented in the figure. Fig. 7.60 shows the equal weighting method and key indexes with their impact on the sustainability index score.

Although theoretically the indexes presented in Fig. 7.60 have equal weights, they vary in their impact on the sustainability index. Repetitively and similar to previous schemes, the economic index turns to be the index that yields the highest sustainability index scores. The other index shows very similar values throughout the changes of their values, which reflect the equal weighting phenomena.

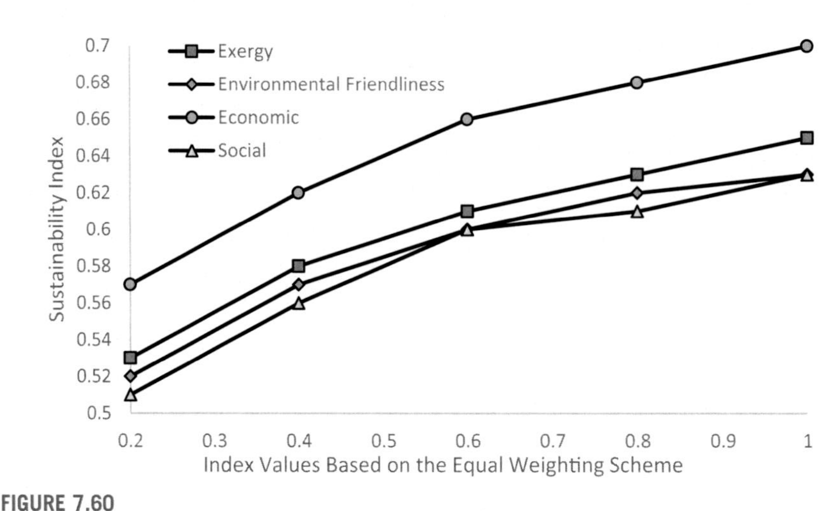

FIGURE 7.60

Variation of the solar PV sustainability index with respect to various indexes based on the equal-weighting scheme.

7.3.6 Closing remarks

In this case study, a comprehensive assessment model has been developed to evaluate the sustainability of energy systems, which is convenient, beneficial to policy makers, and public. The sustainability index for the PV system ranged from 0.74 to 0.73. This integrated sustainability index score is convenient and beneficial to policy makers, researchers, and the public. Moreover, the weighting schemes are shown to have an effect on the final sustainability index with little variations between them. Furthermore, this model introduces new dimensions when assessing sustainability of energy systems. These dimensions proved to be important as they influence the final score. In specific, the sustainability index of the PV system using the hierarchist scheme is 0.79 including all dimensions and 0.7 excluding the introduced dimensions. This means that a system could in fact be more sustainable, but the way the assessment is conducted may alter its results. This study is limited, and further research includes validating this sustainability assessment model on energy systems outside of Ontario. Optimization analysis for the proposed system is suggested for future work as well as applying the model on real energy systems. Lastly, the social impression can be evaluated in a better way. Indicators such as human welfare, human health, social cost, and ethical responsibility can be better assessed to accurately reflect the social impression dimension.

CHAPTER

Future directions and conclusions

8

Sustainability remains to be a very subjective matter and a difficult area of study that can definitely be enhanced further by incorporating various technical and well-established numerical models to achieve less subjectivity. In fact, sustainability is a complex and multidisciplinary concept, which explains the absence of a universally adopted sustainability assessment model until today. This book features a comprehensive understanding about sustainability and sustainable energy solutions. The sustainability assessment model introduced in this book highlights a comprehensive collection of indicators including thermodynamic, environmental, economic, social, and technological to evaluate the sustainability of energy systems. Furthermore, these indicators are dimensionless using target values, which reflect the preferred values under optimal conditions. The idea of using target values to normalize the data is also novel. Moreover, data collected from various indicators are further processed using a variety of aggregation and weighting schemes to minimize the subjectivity associated with the model as much as possible. Using the weighted geometric mean and weighted arithmetic mean along with the panel method, hierarchist, egalitarian, individualist, and equal weighting schemes is novel as the results unveil important information about the behavior and influence of each method on the different categories of data. In addition, the proposed model has been validated using various case studies. Using EES and SimaPro, the case studies are modeled to explore the thermodynamic and environmental impression performance. Renewable energy systems have better environmental performance than conventional fossil-based fuel systems. Wind energy has a better environmental footprint than solar photovoltaic (PV), however both of their environmental performance is not comparable with the environmental impression of the conventional gas-fired system. In summary, the specific conclusions derived from the case studies are as follows:

- Comprehensive assessment model has been developed to evaluate the sustainability of energy systems.
- Integrated sustainability index score is developed, which is convenient and beneficial to policy makers and public.
- Weighting schemes effect the final sustainability index with little variations between them.
- The solar PV system has a low sustainability score of 0.66 using the panel method and a high score of 0.69 using the hierarchist and equal weighting methods based on the weighted geometric mean.

Energy Sustainability. https://doi.org/10.1016/B978-0-12-819556-7.00008-5

- The solar PV system has a low sustainability score of 0.64 using the individualist method and a high score of 0.63 using the equal weighting method based on the weighted arithmetic mean.
- The index with the lowest priority factor according to the panel method is the sizing index with a weight of 0.06. The index with the highest priority factor according to panel method is the environmental friendliness index with a weight of 0.18.
- Increased wind turbine mechanical efficiency has no impact on the wind system's energy efficiency, yet it decreases the exergy efficiency in a linear fashion.
- The wind energy system has a low sustainability score of 0.66 using the panel method and a high score of 0.68 using the hierarchist and equal weighting methods based on the weighted geometric mean.
- The wind energy system has a low sustainability score of 0.61 using the individualist scheme and a high score of 0.6 using the equal weighting scheme based on the weighted arithmetic mean.
- As the wind speed increases, the general trend for the energy and exergy efficiencies is to decrease. Wind speeds less than 22 m/s are characterized with the highest energy and exergy efficiencies.
- The energy and exergy efficiencies of the solar PV system are 66% and 30%, respectively. On the other hand, the energy and exergy efficiencies of the wind system are 31% and 24%, respectively.
- Environmentally, wind energy has the lowest impact on all categories, whereas the solar PV system is rated second.

Although this model introduces novel categories related to the sustainability of energy systems, there is more to do to ensure the maturity of the assessment methodology. The following recommendations are proposed for future sustainability research pertaining to this topic:

- Social index could be measured in a better way. Indicators such as human welfare, human health, social cost, and ethical responsibility can be better assessed to accurately reflect the social impression category.
- Inclusion of more experienced stakeholders in the panel and incorporating other methods of assigning weights such as discussion.
- Although different weighting schemes have been explored when processing the values for the eight domains, equal weighting was adopted throughout all subindicators. This thesis recommends investigating different weight schemes when processing the subindicators.
- Acknowledging the subjectivity of the assessment methodology and the weak parameters such as social indicators and the education indicators, which are hard to qualify; and build on it to minimize errors and weakness of the model.
- Validating this sustainability assessment model on energy systems outside of Ontario. Changing the geographical location may change economic, environmental, and social impacts as well as technical parameters such as wind speed, solar irradiance, or annual household energy demand.

- Carrying out an optimization analysis for the proposed systems and identifying how the system can be enhanced to achieve a better sustainability score.

Furthermore, the future of sustainability research lies in establishing more research centers and think tanks that dedicate their work toward defining and implementing a transformative agenda for applying these sustainability principles. In addition, encouraging greener practices is essential to advance the sustainability agenda forward. Sustainable transportation solutions for smart cities will also play a vital role in this process. In fact, artificial intelligence and machine learning will be critical aspects for the development of smart cities, smart homes, and the enhancement of energy sustainability. Moreover, the future direction for energy systems in light of this revolutionary digital age will revolve around robotics, which will also need internal energy system management.

Nomenclature

Ac	Area of collector, m^2
Cp	Specific heat at constant pressure, kJ/kg K
ex	Specific exergy, kJ/kg
h	Specific enthalpy, kJ/kg
$Impp$	MPP current, A
Isc	Short circuit current, A
I_b	Beam solar radiation on a horizontal surface, W/m^2
I_d	Diffuse solar radiation on a horizontal surface, W/m^2
I_T	Hourly total solar radiation on a horizontal surface, W/m^2
$\dot{m}$	Mass flow rate, kg/s
MATAI	Median after-tax annual income, $/year
MH	Mixing height, m
N_{PVT}	Number of PVT panels
Pop	Population
$\mathbf{Power_{max}}$	Maximum power output, MW
PR	Production rate, tonnes/year
$\dot{Q}$	Heat rate, kW
R	Recoverable reserves, kg
R_a	Solar irradiance, W/m^2
$\dot{S}$	Entropy rate, kW/K
$\dot{S}S$	System size, kW
s	Specific entropy, kJ/kg K
T	Temperature, K
t	Time, year
T_0	Reference temperature, °C
T_{ref}	Reference temperature, °C or K
v	Specific volume, m^3/kg
$Vmpp$	MPP voltage, V
Voc	Open circuit voltage, V
W	Weighting factor
$\dot{W}$	Work rate, kW
$\dot{W}_{pump}$	Pump rate, kW
WAM	Weighted arithmetic mean
WGM	Weighted geometric mean
X	Sustainability indicator
Y	Dimensionless indicator value

Greek Letters

ψ_{geo}	Geothermal system exergy efficiency
β_p	Temperature coefficient for module efficiency
η_{PV}	PV module efficiency
ψ_{PVT}	PVT exergy efficiency
η_{ref}	PV module efficiency at reference temperature
ψ_{system}	System exergy efficiency

α	Adjustment factor
η	Energy efficiency
τ	Residence time, h
ψ	Exergy efficiency

Subscript

amb	Ambient
Comb	Combustion
Cond	Condenser
D	Destruction
ED	Exergy destruction
ef	Efficiency of the system, %
ER	Energy
EX	Exergy
ef(T)	Target efficiency, %
ENV	Environment
Evap	Evaporator
0	Reference environment
Sust	Sustainability
(T)	Target value

Abbreviations

ADP	Abiotic depletion potential
AP	Acidification potential
AT	Air toxicity
BCR	Benefit-cost ratio
CFC	Chlorofluorocarbon
COMM	Commercializability
DCB	Dichlorobenzene
DHW	Domestic hot water
EES	Engineering equation solver
EFI	Environmental friendliness index
EI	Educational innovation
EL	Educational level
EP	Eutrophication potential
EPA	Environmental protection agency
GHG	Greenhouse gases
GWP	Global warming potential
HH	Human health
HW	Human welfare
IN	Innovation
JC	Job creation
LCA	Lifecycle assessment
LCOE	Levelized cost of electricity/energy
LU	Land use
microFIT	Micro feed in tariff

NPV	Net present value
ODP	Ozone depletion potential
PA	Public awareness
PBT	Payback time
PM	Particulate matter
PV	Photovoltaics
SA	Social acceptance
SC	Social cost
TR	Technology readiness
TRAIN	Training
WC	Water consumption
WE	Water ecotoxicity
IESO	Independent electricity service operator
NERL	National exposure research laboratory
ROI	Return on investment
SAM	System advisor model

Bibliography

Abu-Rayash, A., Dincer, I., 2019. Sustainability assessment of energy systems: a novel integrated model. Journal of Cleaner Production 212, 1098–1116. https://doi.org/10.1016/J.JCLEPRO.2018.12.090.

Afgan, N.H., Carvalho, M.G., 2004. Sustainability assessment of hydrogen energy systems. International Journal of Hydrogen Energy 29 (13), 1327–1342. https://doi.org/10.1016/J.IJHYDENE.2004.01.005.

Akter, M.N., Mahmud, M.A., Oo, A.M.T., 2017. Comprehensive economic evaluations of a residential building with solar photovoltaic and battery energy storage systems: an Australian case study. Energy and Buildings 138, 332–346. https://doi.org/10.1016/J.ENBUILD.2016.12.065.

AlFaris, F., Juaidi, A., Manzano-Agugliaro, F., 2017. Intelligent homes' technologies to optimize the energy performance for the net zero energy home. Energy and Buildings 153, 262–274. https://doi.org/10.1016/J.ENBUILD.2017.07.089.

Antonelli, M., Desideri, U., 2014. The doping effect of Italian feed-in tariffs on the PV market. Energy Policy 67, 583–594. https://doi.org/10.1016/J.ENPOL.2013.12.025.

Axaopoulos, P.J., Fylladitakis, E.D., 2013. Performance and economic evaluation of a hybrid photovoltaic/thermal solar system for residential applications. Energy and Buildings 65, 488–496. https://doi.org/10.1016/J.ENBUILD.2013.06.027.

Ayoub, J., 2013. Towards net zero energy solar buildings. CanmetEnergy, Natural Resources Canada, Government of Canada.

Babu, P., Kumar, R., Linga, P., 2013. Pre-combustion capture of carbon dioxide in a fixed bed reactor using the clathrate hydrate process. Energy 50, 364–373. https://doi.org/10.1016/j.energy.2012.10.046.

Balfaqih, H., Al-Nory, M.T., Nopiah, Z.M., Saibani, N., 2017. Environmental and economic performance assessment of desalination supply chain. Desalination 406, 2–9. https://doi.org/10.1016/J.DESAL.2016.08.004.

Bertani, R., 2005. World geothermal power generation in the period 2001–2005. Geothermics 34 (6), 651–690. https://doi.org/10.1016/j.geothermics.2005.09.005.

Bicer, Y., Dincer, I., 2016. Analysis and performance evaluation of a renewable energy based multigeneration system. Energy 94, 623–632. https://doi.org/10.1016/J.ENERGY.2015.10.142.

Bloemhof, B., 2017. Assessing consumer benefits in the Ontario residential retail natural gas market: Ontario residential retail natural gas market: why marketer entry did not help. Energy Policy 109, 555–564. https://doi.org/10.1016/J.ENPOL.2017.07.038.

Bloomberg, N.E.F., 2018. Global Electricity Demand to Increase 57% by 2050, 2018, September 04) Retrieved from. https://about.bnef.com/blog/global-electricity-demand-increase-57-2050/.

Booysen, F., 2002. An overview and evaluation of composite indices of development, 59 (2), 115–151.

Boza-Kiss, B., Moles-Grueso, S., Urge-Vorsatz, D., 2013. Evaluating policy instruments to foster energy efficiency for the sustainable transformation of buildings. Current Opinion in Environmental Sustainability 5 (2), 163–176. https://doi.org/10.1016/J.COSUST.2013.04.002.

BP Statistical Review of World Energy, 2017. BP, London.

Bragança, L., Mateus, R., Koukkari, H., 2010. Building sustainability assessment. Sustainability 2 (7), 2010–2023. https://doi.org/10.3390/su2072010.

Caliskan, H., Dincer, I., Hepbasli, A., 2011. Exergoeconomic, enviroeconomic and sustainability analyses of a novel air cooler. Energy and Buildings 55, 747–756. https://doi.org/10.1016/j.enbuild.2012.03.024.

Caliskan, H., Dincer, I., Hepbasli, A., 2012. Exergoeconomic, enviroeconomic and sustainability analyses of a novel air cooler. Energy and Buildings 55, 747–756. https://doi.org/10.1016/J.ENBUILD.2012.03.024.

Caliskan, H., Dincer, I., Hepbasli, A., 2013. Energy, exergy and sustainability analyses of hybrid renewable energy based hydrogen and electricity production and storage systems: modeling and case study. Applied Thermal Engineering 61 (2), 784–798. https://doi.org/10.1016/J.APPLTHERMALENG.2012.04.026.

Çengel, Y.A., Boles, M.A., Kanŏglu, M., 2010. Thermodynamics: an engineering approach, seventh ed. McGraw-Hill Education, New York, NY.

Çengel, Y.A., Boles, M.A., Kanŏglu, M., 2019. Thermodynamics: An engineering approach. McGraw-Hill, New York.

Chandrasena, R.P., Shahnia, F., Ghosh, A., Rajakaruna, S., 2014. Secondary control in microgrids for dynamic power sharing and voltage/frequency adjustment. 2014 Australasian Universities Power Engineering Conference (AUPEC), 1–8.

Chapman, A.J., McLellan, B., Tezuka, T., 2016. Residential solar PV policy: an analysis of impacts, successes and failures in the Australian case. Renewable Energy 86, 1265–1279. https://doi.org/10.1016/J.RENENE.2015.09.061.

Chong, Z.R., Yang, S.H.B., Babu, P., Linga, P., Li, X.-S., 2016. Review of natural gas hydrates as an energy resource: prospects and challenges. Applied Energy 162, 1633–1652. https://doi.org/10.1016/J.APENERGY.2014.12.061.

Clarke, K.D., Scruton, D.A., 1997. The benthic community of stream riffles in Newfoundland, Canada and its relationship to selected physical and chemical parameters. Journal of Freshwater Ecology 12 (1), 113–121. https://doi.org/10.1080/02705060.1997.9663514.

Cronemberger, J., Caamaño-Martín, E., Sánchez, S.V., 2012. Assessing the solar irradiation potential for solar photovoltaic applications in buildings at low latitudes – making the case for Brazil. Energy and Buildings 55, 264–272. https://doi.org/10.1016/J.ENBUILD.2012.08.044.

CSD, 2001. Indicators of Sustainable Development: Guidelines and Methodologies. Commission on Sustainable Development, New York, USA. http://www.un.org/esa/sustdev/natlinfo/indicators/indisd/indisd-mg2001.pdf.

Dampier, J.E.E., Shahi, C., Lemelin, R.H., Luckai, N., 2016. Assessment of potential local and regional induced economic impact of an energy policy change in rural northwestern Ontario. Energy, Sustainability and Society 6, 1–11.

De Vries, J.W., Vinken, T.M.W.J., Hamelin, L., De Boer, I.J.M., 2012. Comparing environmental consequences of anaerobic mono- and co-digestion of pig manure to produce bio-energy – a life cycle perspective. Bioresource Technology 125, 239–248. https://doi.org/10.1016/J.BIORTECH.2012.08.124.

Delisle, V., 2011. Net-zero energy houses: Solar photovoltaic electricity scenario analysis based on current and future costs. ASHRAE Transactions 112 (2), 315–322.

Dincer, I., 2000. Renewable energy and sustainable development: a crucial review. Renewable and Sustainable Energy Reviews 4 (2), 157–175. https://doi.org/10.1016/S1364-0321(99)00011-8.

Dincer, I., 2007. Exergetic and sustainability aspects of green energy systems. Clean – Soil, Air, Water 36 (4), 311–322. https://doi.org/10.1002/clen.200700108.

Dincer, I., 2016. Exergization. International Journal of Energy Research 40 (14), 1887–1889. https://doi.org/10.1002/er.3606.

Dincer, I., Acar, C., 2015. Review and evaluation of hydrogen production methods for better sustainability. International Journal of Hydrogen Energy 40 (34), 11094–11111. https://doi.org/10.1016/j.ijhydene.2014.12.035.

Dincer, I., Acar, C., 2016. A review on clean energy solutions for better sustainability. International Journal of Energy Research 39 (6), 585–606. https://doi.org/10.1002/er.3329.

Dincer, I., Acar, C., 2017a. Innovation in hydrogen production. International Journal of Hydrogen Energy 42 (22), 14843–14864. https://doi.org/10.1016/j.ijhydene.2017.04.107.

Dincer, I., Acar, C., 2017b. Smart energy systems for a sustainable future. Applied Energy 194, 225–235. https://doi.org/10.1016/j.apenergy.2016.12.058.

Dincer, I., Acar, C., 2018. Smart energy solutions with hydrogen options. International Journal of Hydrogen Energy 43 (18), 8579–8599. https://doi.org/10.1016/j.ijhydene.2018.03.120.

Dincer, I., Rosen, M.A., 2004. Exergy as a driver for achieving sustainability. International Journal Of Green Energy 1 (1), 1–19. https://doi.org/10.1081/ge-120027881.

Dincer, I., Rosen, M., 2007. Energetic, exergetic, environmental and sustainability aspects of thermal energy storage systems. NATO Science Series Thermal Energy Storage for Sustainable Energy Consumption 23–46. https://doi.org/10.1007/978-1-4020-6290-3_2.

Dincer, I., Rosen, M.A., 2011. Sustainability aspects of hydrogen and fuel cell systems. Energy for Sustainable Development 15 (2), 137–146. https://doi.org/10.1016/J.ESD.2011.03.006.

Dincer, I., Zamfirescu, C., 2012. Potential options to greenize energy systems. Energy 46 (1), 5–15. https://doi.org/10.1016/J.ENERGY.2011.11.061.

Diouf, B., Pode, R., 2015. Potential of lithium-ion batteries in renewable energy. Renewable Energy 76, 375–380. https://doi.org/10.1016/J.RENENE.2014.11.058.

Dong, C., Wiser, R., Rai, V., 2018. Incentive pass-through for residential solar systems in California. Energy Economics 72, 154–165. https://doi.org/10.1016/J.ENECO.2018.04.014.

Duffie, J.A., Beckman, W.A., 2013. Solar Engineering of Thermal Processes. John Wiley & Sons.

ECO., 2017. Every Joule Counts (Rep.). Toronto, ON: Environmental Commissioner of Ontario. 2, 17–23.

Environment and Climate Change Canada, 2017. National Inventory Report 1990–2016: Greenhouse Gas Sources and Sinks in Canada, Ottawa.

EPA, 2011. National Ambient Air Quality Standards. Washington, D.C. Available online. http://www.epa.gov/air/criteria.

Evans, D.L., 1981. Simplified method for predicting photovoltaic array output. Solar Energy 27 (6), 555–560. https://doi.org/10.1016/0038-092X(81)90051-7.

Evans, A., Strezov, V., Evans, T.J., 2009. Assessment of sustainability indicators for renewable energy technologies. Renewable and Sustainable Energy Reviews 13 (5), 1082–1088. https://doi.org/10.1016/J.RSER.2008.03.008.

Evrendilek, F., Ertekin, C., 2003. Assessing the potential of renewable energy sources in Turkey. Renewable Energy 28 (15), 2303–2315. https://doi.org/10.1016/S0960-1481(03)00138-1.

Fremeth, A., Holburn, G., Loudermilk, M., Schaufele, B., 2017. The Economic Cost of Electricity Generation in Ontario, 1–18. The Ivey Energy Policy and Management Centre.

Gagnon, L., Belanger, C., Uchiyama, Y., 2002. Life-cycle assessment of electricity gen-eration options: the status of research in year 2001. Energy Policy 30, 1267–1278.

Ghimire, A., Frunzo, L., Pirozzi, F., Trably, E., Escudie, R., Lens, P.N.L., Esposito, G., 2015. A review on dark fermentative biohydrogen production from organic biomass: process parameters and use of by-products. Applied Energy 144, 73–95. https://doi.org/10.1016/J.APENERGY.2015.01.045.

Gnanapragasam, N.V., Reddy, B.V., Rosen, M.A., 2010. A methodology for assessing the sustainability of hydrogen production from solid fuels. Sustainability 2 (6), 1472–1491. https://doi.org/10.3390/su2061472.

Godfrey, L., Todd, C., 2001. Defining thresholds for freshwater sustainability indicators within the context of South African water resource management. In: 2nd WARFA/Waternet Symposium: Integrated Water Resource Management: Theory, Practice, Cases, Cape Town, South Africa, http://www.waternetonline.ihe.nl/aboutWN/pdf/godfrey.pdf.

Goedkoop, M., Spriensma, R., 2000. The Eco-Indicator 99: A damage oriented method for life cycle impact assessment. PRé Consultants, Amersfoort, Netherlands.

Götz, M., Lefebvre, J., Mörs, F., McDaniel Koch, A., Graf, F., Bajohr, S., Kolb, T., 2016. Renewable power-to-gas: a technological and economic review. Renewable Energy 85, 1371–1390. https://doi.org/10.1016/J.RENENE.2015.07.066.

Govindan, K., 2015. Green sourcing: Taking steps to achieve sustainability management and conservation of resources. Resources, Conservation and Recycling 104, 329–333.

Govindan, K., 2016. Green sourcing: taking steps to achieve sustainability management and conservation of resources. Resources, Conservation and Recycling 104, 329–333.

GRI – Global Reporting Initiative 2002a, 2004. The Global Reporting Initiative – An Overview. Global Reporting Initiative, Boston, USA. Available at. http://www.globalreporting.org.

Hacatoglu, K., 2014. A Systems Approach to Assessing the Sustainability of Hybrid Community Energy Systems (PhD's thesis). University of Ontario Institute of Technology, Oshawa.

Hacatoglu, K., Dincer, I., Rosen, M.A., 2015. Sustainability assessment of a hybrid energy system with hydrogen-based storage. International Journal of Hydrogen Energy 40 (3), 1559–1568. https://doi.org/10.1016/J.IJHYDENE.2014.11.079.

Hacatoglu, K., Dincer, I., Rosen, M.A., 2016. Sustainability of a wind-hydrogen energy system: Assessment using a novel index and comparison to a conventional gas-fired system. International Journal of Hydrogen Energy 41 (19), 8376–8385. https://doi.org/10.1016/j.ijhydene.2016.01.135.

Hacatoglu, K., Dincer, I., Rosen, M., 2016. A new model to assess the environmental impact and sustainability of energy systems. Journal of Cleaner Production 103, 211–218. https://doi.org/10.1016/j.jclepro.2014.06.060.

Harwood, Richard R., 1990. A History of Sustainable Agriculture. In: Edwards, Clive A., et al. (Eds.), Sustainable agricultural systems. Soil and Conservation Society.

Holden, M., Fine, B., Rustomjee, Z., 1997. The political economy of South Africa from minerals-energy complex to industrialization. Foreign Policy 108, 163. https://doi.org/10.2307/1149102.

Homagain, K., Shahi, C., Luckai, N., Sharma, M., 2016. Life cycle cost and economic assessment of biochar-based bioenergy production and biochar land application in Northwestern Ontario, Canada. Forest Ecosystems 3 (1), 21. https://doi.org/10.1186/s40663-016-0081-8.

Hoppmann, J., 2015. The role of deployment policies in fostering innovation for clean energy technologies: insights from the solar photovoltaic industry. Business and Society 54 (4), 540–558. https://doi.org/10.1177/0007650314558042.

IAEA Annual Report, 2017. International Atomic Energy Agency, Vienna.
IEA., 2018. Key World Energy Statistics. https://doi.org/10.1787/key_energ_stat-2018-en.
IESO, 2016. Ontario Energy Report Q1 2016 (Rep.). Independent Electricity Service Operator, Toronto, ON, 1–16.
IESO, 2016. Ontario Energy Report Q1 2016. Independent Electricity Service Operator, Toronto, ON, pp. 1–16.
Inhaber, H., 2004. Water use in renewable and conventional electricity production. Energy Sources 26 (3), 309–322. https://doi.org/10.1080/00908310490266698.
Ioannou, A.K., Stefanakis, N.E., Boudouvis, A.G., 2014. Design optimization of residential grid-connected photovoltaics on rooftops. Energy and Buildings 76, 588–596. https://doi.org/10.1016/J.ENBUILD.2014.03.019.
IRENA and CEM, 2014. The socio-economic benefits of large-scale solar and wind: an econValue report.
IRG, 2012. National Nuclear Attitude Survey (Rep.). Innovative Research Group, Toronto, ON, pp. 1–37.
Jin, E., Mendis, G.P., Sutherland, J.W., 2019. Integrated sustainability assessment for a bioenergy system: A system dynamics model of switchgrass for cellulosic ethanol production in the U.S. midwest. Journal of Cleaner Production 234, 503–520. https://doi.org/10.1016/j.jclepro.2019.06.205.
Juwana, I., Muttil, N., Perera, B.J.C., 2012. Indicator-based water sustainability assessment — a review. The Science of the Total Environment 438, 357–371. https://doi.org/10.1016/J.SCITOTENV.2012.08.093.
Kahn Ribeiro, S., Roy, J., Urge-Vorsatz, D., Figueroa, M., 2013. Energy systems in the context of sustainable development. Current Opinion in Environmental Sustainability 5 (2), 136–140.
Kalinci, Y., Hepbasli, A., 2015. Techno-economic analysis of a stand-alone hybrid renewable energy system with hydrogen production and storage options. International Journal of Hydrogen Energy 40 (24), 7652–7664. https://doi.org/10.1016/J.IJHYDENE.2014.10.147.
Kammen, D.M., Pacca, S., 2004. Assessing the costs of electricity. Annual Review of Environment and Resources 29 (1), 301–344. https://doi.org/10.1146/annurev.energy.28.050302.105630.
Kanoglu, M., Dincer, I., Cengel, Y.A., 2008. Exergy for better environment and sustainability. Environment, Development and Sustainability 11 (5), 971–988. https://doi.org/10.1007/s10668-008-9162-3.
Kates, R.W., Clark, W.C., Corell, R., Hall, J.M., Jaeger, C.C., Lowe, I., et al., 2001. Sustainability science. Science 292 (5517), 641 LP-642. https://doi.org/10.1126/science.1059386.
Kato, K., Murata, A., Sakuta, K., 1998. Energy pay-back time and life-cycle CO2 emission of residential PV power system with silicon PV module. Progress in Photovoltaics: Research and Applications 6 (2), 105–115.
Kucuk, H., Midilli, A., Özdemir, A., Çakmak, E., Dincer, I., 2010. Exergetic performance analysis of a recirculating aquaculture system. Energy Conversion and Management 51 (5), 1033–1043. https://doi.org/10.1016/J.ENCONMAN.2009.12.007.
Lackner, K.S., Sachs, J., 2005. A robust strategy for sustainable energy. Brookings Papers on Economic Activity (2), 215–284. https://doi.org/10.1353/eca.2006.0007.

Lamnatou, Chr., Chemisana, D., 2017. Concentrating solar systems: Life Cycle Assessment (LCA) and environmental issues. Renewable and Sustainable Energy Reviews 78 (issue C), 916–932. https://EconPapers.repec.org/RePEc:eee:rensus:v:78:y:2017:i:c.

Lancker, E., Nijkamp, P., 2000. A policy scenario analysis of sustainable agricultural development options: a case study for Nepal. Impact Assessment and Project Appraisal 18 (2), 111–124. https://doi.org/10.3152/147154600781767493.

Lazard, 2014. Levelized Cost of Energy Analysis, eighth ed. Rep. Retrieved from. https://www.lazard.com/media/1777/levelized_cost_of_energy_-_version_80.pdf.

Leckner, M., Zmeureanu, R., 2011. Life cycle cost and energy analysis of a net zero energy house with solar combisystem. Applied Energy 88 (1), 232–241. https://doi.org/10.1016/J.APENERGY.2010.07.031.

Lenzen, M., 2008. Life cycle energy and greenhouse gas emissions of nuclear energy: a review. Energy Conversion and Management 49 (8), 2178–2199. https://doi.org/10.1016/J.ENCONMAN.2008.01.033.

Lo Piano, S., Mayumi, K., 2017. Toward an integrated assessment of the performance of photovoltaic power stations for electricity generation. Applied Energy 186, 167–174. https://doi.org/10.1016/J.APENERGY.2016.05.102.

Luo, X., Wang, J., Dooner, M., Clarke, J., 2015. Overview of current development in electrical energy storage technologies and the application potential in power system operation. Applied Energy 137, 511–536. https://doi.org/10.1016/J.APENERGY.2014.09.081.

Mallia, E., Lewis, G., 2013. Life cycle greenhouse gas emissions of electricity generation in the province of Ontario, Canada. The International Journal of Life Cycle Assessment 18 (2), 377–391.

Martire, S., Castellani, V., Sala, S., 2015. Carrying capacity assessment of forest resources: Enhancing environmental sustainability in energy production at local scale. Resources, Conservation and Recycling 94, 11–20. https://doi.org/10.1016/J.RESCONREC.2014.11.002.

Mathiesen, B.V., Lund, H., Connolly, D., Wenzel, H., Østergaard, P.A., Möller, B., et al., 2015. Smart energy systems for coherent 100% renewable energy and transport solutions. Applied Energy 145, 139–154. https://doi.org/10.1016/J.APENERGY.2015.01.075.

Mclaughlin, J., Smith, M., 2002. Power of the people: Public awareness and involvement in renewable energy in Scotland. Refocus 3 (5), 26–29. https://doi.org/10.1016/s1471-0846(02)80082-8.

Midilli, A., Kucuk, H., Dincer, I., 2012. Environmental and sustainability aspects of a recirculating aquaculture system. Environmental Progress and Sustainable Energy 31 (4), 604–611. https://doi.org/10.1002/ep.10580.

Ministry of Energy, 2017. The 2017 Long-Term Energy Plan (Rep.). Ministry of Energy, Toronto, ON.

Modarres, A., 2017. Commuting, energy consumption, and the challenge of sustainable urban development. Current Opinion in Environmental Sustainability 25, 1–7. https://doi.org/10.1016/J.COSUST.2017.01.011.

Morelli, John, 2011. Environmental sustainability: A definition for environmental professionals. Journal of Environmental Sustainability Vol. 1 (Iss. 1). Article 2. https://doi.org/10.14448/jes.01.0002.

Munda, G., 2005. "Measuring Sustainability": A multi-criterion framework. Environment, Development and Sustainability 7 (1), 117–134. https://doi.org/10.1007/s10668-003-4713-0.

Nastasi, B., Lo Basso, G., 2016. Hydrogen to link heat and electricity in the transition towards future smart energy systems. Energy 110, 5–22. https://doi.org/10.1016/J.ENERGY.2016.03.097.

National Energy Board, 2016. Canada's Energy Future 2016 Update. NEB, pp. 1–48. Rep. ISSN. 2369-1479).

National Energy Board, 2017. Canada's Adoption of Renewable Power Sources. NEB, pp. 1–32. ISSN. No. 2371-6804).

Natural Resources Canada., 2017. Residential End-Use Model, Ottawa.

Natural Resources Canada, 2017a. Commercial/Institutional End-Use Model, Ottawa.

Natural Resources Canada, 2017b. Residential End-Use Model, Ottawa.

NEB, 2017a. Canada's Energy Future 2017: Energy Supply and Demand Projections to 2040. National Energy Board, pp. 1–90.

NEB, 2017b. Provincial and Territorial Energy Profiles – Ontario (Rep.). National Energy Board, Ottawa, ON.

NEB., 2017. Provincial and Territorial Energy Profiles - Ontario (Rep.). National Energy Board, Ottawa, ON.

NEB., 2017. Canada's Energy Future 2017: Energy Supply and Demand Projections to 2040 (Tech.). National Energy Board, 1–90.

Ness, B., Urbel-Piirsalu, E., Anderberg, S., Olsson, L., 2007. Categorising tools for sustainability assessment. Ecological Economics 60 (3), 498–508. https://doi.org/10.1016/J.ECOLECON.2006.07.023.

Newell, B., 2014a. Net zero houses. ASHRAE Journal 38.

Newell, B., 2014b. Net zero homes. ASHRAE Journal 56 (5), 38. Academic OneFile.

Ning, G., Junnan, L., Yansong, D., Zhifeng, Q., Qingshan, J., Weihua, G., Geert, D., 2017. BIM-based PV system optimization and deployment. Energy and Buildings 150, 13–22. https://doi.org/10.1016/J.ENBUILD.2017.05.082.

OEB, 2016. Natural Gas and Electricity Utility Yearbooks (Tech.). Ontario Energy Board, Toronto, ON.

OEB., 2016. Natural gas and electricity utility yearbooks (Tech.). Ontario Energy Board, Toronto, ON.

OECD – Organization for Economic Co-operation and Development, 2004. An Update of the OECD Composite Leading Indicators. Short-term Economic Statistics Division, Statistics Directorate/OECD. Available at. http://www.oecd.org.

Ontario Ministry of Energy, 2017. Ontario's Long Term Energy Plan 2017. Ministry of Energy.

Ontario Ministry of Infrastructure, 2014. Energy Consumption and Greenhouse Gas Emission Report (Rep.). Toronto, ON: Retrieved from. https://www.ontario.ca/page/2014-energy-consumption-and-greenhouse-gas-emission-report#section-0.

Ozbilen, A., Dincer, I., Naterer, G.F., Aydin, M., 2012. Role of hydrogen storage in renewable energy management for Ontario. International Journal of Hydrogen Energy 37 (9), 7343–7354. https://doi.org/10.1016/J.IJHYDENE.2012.01.073.

Paolotti, L., Martino, G., Marchini, A., Boggia, A., 2017. Economic and environmental assessment of agro-energy wood biomass supply chains. Biomass and Bioenergy 97, 172–185. https://doi.org/10.1016/J.BIOMBIOE.2016.12.020.

Pearce, D.W., Markandya, A., 1989. The Benefits of Environmental Policy: Monetary Valuation.

Perez, R., Ineichen, P., Seals, R., Michalsky, J., Stewart, R., 1990. Modeling daylight availability and irradiance components from direct and global irradiance. Solar Energy 44 (5), 271–289. https://doi.org/10.1016/0038-092X(90)90055-H.

Permpool, N., Gheewala, S.H., 2017. Environmental and energy assessment of alternative fuels for diesel in Thailand. Journal of Cleaner Production 142, 1176–1182. https://doi.org/10.1016/j.jclepro.2016.08.081.

Piano, S.L., Mayumi, K., 2017. Toward an integrated assessment of the performance of photovoltaic power stations for electricity generation. Applied Energy 186, 167–174. https://doi.org/10.1016/j.apenergy.2016.05.102.

Pierie, F., Bekkering, J., Benders, R., Gemert, W.V., Moll, H., 2016. A new approach for measuring the environmental sustainability of renewable energy production systems: Focused on the modelling of green gas production pathways. Applied Energy 162, 131–138. https://doi.org/10.1016/j.apenergy.2015.10.037.

Pirnia, M., Nathwani, J., Fuller, D., 2011. Ontario feed-in-tariffs: system planning implications and impacts on social welfare. The Electricity Journal 24 (8), 18–28. https://doi.org/10.1016/J.TEJ.2011.09.009.

Polzin, F., Sanders, M., Täube, F., 2017. A diverse and resilient financial system for investments in the energy transition. Current Opinion in Environmental Sustainability 28, 24–32. https://doi.org/10.1016/J.COSUST.2017.07.004.

Ramachandran, N., 2000. Monitoring Sustainability: Indices and Techniques of Analysis. Concept Publishing Company.

Ricotti, M.E., 2013. Nuclear energy: basics, present, future. EPJ Web of Conferences 54, 1005. https://doi.org/10.1051/epjconf/20135401005.

Rowley, H.V., Peters, G.M., Lundie, S., Moore, S.J., 2012. Aggregating sustainability indicators: beyond the weighted sum. Journal of Environmental Management 111, 24–33. https://doi.org/10.1016/J.JENVMAN.2012.05.004.

Rutovitz, J., Dominish, E., Downes, J., 2015. Calculating global energy sector jobs: 2015 methodology. Prepared for Greenpeace International by the Institute for Sustainable Futures, University of Technology, Sydney.

Sagani, A., Mihelis, J., Dedoussis, V., 2017. Techno-economic analysis and life-cycle environmental impacts of small-scale building-integrated PV systems in Greece. Energy and Buildings 139, 277–290. https://doi.org/10.1016/J.ENBUILD.2017.01.022.

Saldanha, N., Beausoleil-Morrison, I., 2012. Measured end-use electric load profiles for 12 Canadian houses at high temporal resolution. Energy and Buildings 49, 519–530. https://doi.org/10.1016/J.ENBUILD.2012.02.050.

Sandler, S.I., 1999. Chemical and Engineering Thermodynamics, third ed. John Wiley and Sons, New York.

Šantek, B., Gwehenberger, G., Šantek, M.I., Narodoslawsky, M., Horvat, P., 2010. Evaluation of energy demand and the sustainability of different bioethanol production processes from sugar beet. Resources, Conservation and Recycling 54 (11), 872–877.

Santoyo-Castelazo, E., Azapagic, A., 2014. Sustainability assessment of energy systems: integrating environmental, economic and social aspects. Journal of Cleaner Production 80, 119–138. https://doi.org/10.1016/J.JCLEPRO.2014.05.061.

Sarbu, I., Sebarchievici, C., 2018. A Comprehensive Review of Thermal Energy Storage. Sustainability 10 (2), 191. https://doi.org/10.3390/su10010191.

Sardianou, E., Genoudi, P., 2013. Which factors affect the willingness of consumers to adopt renewable energies? Renewable Energy 57, 1–4. https://doi.org/10.1016/J.RENENE.2013.01.031.

Scholes, G.D., 2003. Long-range resonance energy transfer in molecular systems. Annual Review of Physical Chemistry 54, 57–87. Retrieved from. http://search.proquest.com.uproxy.library.dc-uoit.ca/docview/73225471?accountid=14694.

SEC., 2013. The Economic Impacts of Reducing Natural Gas and Electricity Use in Ontario (Tech.). Stokes Economic Consulting Inc., Milton, ON, pp. 1–18.

Simpson, G., Clifton, J., 2015. The emperor and the cowboys: the role of government policy and industry in the adoption of domestic solar microgeneration systems. Energy Policy 81, 141–151. https://doi.org/10.1016/J.ENPOL.2015.02.028.

Singh, R.K., Murty, H.R., Gupta, S.K., Dikshit, A.K., 2012. An overview of sustainability assessment methodologies. Ecological Indicators 15 (1), 281–299. https://doi.org/10.1016/J.ECOLIND.2011.01.007.

Sleeswijk, A.W., van Oers, L.F.C.M., Guinée, J.B., Struijs, J., Huijbregts, M.A.J., 2008. Normalisation in product life cycle assessment: an LCA of the global and European economic systems in the year 2000. The Science of the Total Environment 390 (1), 227–240. https://doi.org/10.1016/J.SCITOTENV.2007.09.040.

Sørensen, B., 2011. Life-cycle Analysis of Energy Systems: From Methodology to Applications. Royal Society of Chemistry.

Spang, E.S., Moomaw, W.R., Gallagher, K.S., Kirshen, P.H., Marks, D.H., 2014. The water consumption of energy production: an international comparison. Environmental Research Letters 9 (10), 105002. https://doi.org/10.1088/1748-9326/9/10/105002.

Spertino, F., Ciocia, A., Di Leo, P., Tommasini, R., Berardone, I., Corrado, M., Paggi, M., 2015. A power and energy procedure in operating photovoltaic systems to quantify the losses according to the causes. Solar Energy 118, 313–326. https://doi.org/10.1016/J.SOLENER.2015.05.033.

Statistics Canada, 2017. Report on Energy Supply and Demand in Canada, 1990-2016. CANSIM (Table 128-0016), Ottawa, 2017.

Sutherland, J.W., Richter, J.S., Hutchins, M.J., Dornfeld, D., Dzombak, R., Mangold, J., et al., 2016. The role of manufacturing in affecting the social dimension of sustainability. CIRP Annals 65 (2), 689–712. https://doi.org/10.1016/J.CIRP.2016.05.003.

Szekeres, A., Jeswiet, J., 2018. Effects of technological development and electricity price reductions on adoption of residential heat pumps in Ontario, Canada. International Journal of Energy and Environmental Engineering 1–13.

Szekeres, A., Jeswiet, J., 2018. Effects of technological development and electricity price reductions on adoption of residential heat pumps in Ontario, Canada. International Journal of Energy and Environmental Engineering 9 (2), 201–213. https://doi.org/10.1007/s40095-018-0263-y.

Thomas, W.D., Duffy, J.J., 2013. Energy performance of net-zero and near net-zero energy homes in New England. Energy and Buildings 67, 551–558. https://doi.org/10.1016/J.ENBUILD.2013.08.047.

Tirado Herrero, S., Nicholls, L., Strengers, Y., 2018. Smart home technologies in everyday life: do they address key energy challenges in households? Current Opinion in Environmental Sustainability 31, 65–70.

Tirado Herrero, S., Nicholls, L., Strengers, Y., 2018. Smart home technologies in everyday life: do they address key energy challenges in households? Current Opinion in Environmental Sustainability 31, 65–70. https://doi.org/10.1016/J.COSUST.2017.12.001.

Tongsopit, S., 2015. Thailand's feed-in tariff for residential rooftop solar PV systems: progress so far. Energy for Sustainable Development 29, 127–134. https://doi.org/10.1016/J.ESD.2015.10.012.

Ulubeyli, S., Kazanci, O., 2018. Holistic sustainability assessment of green building industry in Turkey. Journal of Cleaner Production 202, 197–212. https://doi.org/10.1016/J.JCLEPRO.2018.08.111.

US Department of Energy, 2011. Office of Fossil Energy and Office of Policy and International Affairs. "Refined Petroleum Product Reserve: Assessment of Energy Security Needs, Costs and Benefits. " September 2011.

Wang, Z., Wennersten, R., Sun, Q., 2017a. Outline of principles for building scenarios – transition toward more sustainable energy systems. Applied Energy 185, 1890–1898. https://doi.org/10.1016/J.APENERGY.2015.12.062.

Wang, Z., Xu, G., Ren, J., Li, Z., Zhang, B., Ren, X., 2017. Polygeneration system and sustainability: Multi-attribute decision-support framework for comprehensive assessment under uncertainties. Journal of Cleaner Production 167, 1122–1137.

Wang, Z., Xu, G., Ren, J., Li, Z., Zhang, B., Ren, X., 2017b. Polygeneration system and sustainability: multi-attribute decision-support framework for comprehensive assessment under uncertainties. Journal of Cleaner Production 167, 1122–1137. https://doi.org/10.1016/J.JCLEPRO.2017.08.089.

WBCSD – World Business Council for Sustainable Development, 1997. Signals of Change: Business Progress Toward sustainable Development. Geneva, Switzerland.

Weitemeyer, S., Kleinhans, D., Vogt, T., Agert, C., 2015. Integration of renewable energy sources in future power systems: the role of storage. Renewable Energy 75, 14–20. https://doi.org/10.1016/J.RENENE.2014.09.028.

White, W., Lunnan, A., Nybakk, E., Kulisic, B., 2013. The role of governments in renewable energy: the importance of policy consistency. Biomass and Bioenergy 57, 97–105. https://doi.org/10.1016/J.BIOMBIOE.2012.12.035.

Winfield, Mark, Dolter, Brett, 2014. Energy, economic and environmental discourses and their policy impact: The case of Ontario's Green Energy and Green Economy Act. Energy Policy 68, 423–435. https://doi.org/10.1016/j.enpol.2014.01.039.

World Nuclear Association, 2018. World Nuclear Performance Report 2018 (Rep. No. 2018/004). England and Wales, London.

Yi, H., Srinivasan, R.S., Braham, W.W., Tilley, D.R., 2017. An ecological understanding of net-zero energy building: evaluation of sustainability based on emergy theory. Journal of Cleaner Production 143, 654–671. https://doi.org/10.1016/J.JCLEPRO.2016.12.059.

Zamfirescu, C., Dincer, I., Stern, M., Wagar, W.R., 2012. Exergetic, environmental and economic analyses of small-capacity concentrated solar-driven heat engines for power and heat cogeneration. International Journal of Energy Research 36 (3), 397–408. https://doi.org/10.1002/er.1811.

Zhang, D., Chai, Q., Zhang, X., He, J., Yue, L., Dong, X., Wu, S., 2012. Economical assessment of large-scale photovoltaic power development in China. Energy 40 (1), 370–375. https://doi.org/10.1016/j.energy.2012.01.053.

Zhang, P., Lv, F.Y., 2015. A review of the recent advances in superhydrophobic surfaces and the emerging energy-related applications. Energy 82, 1068–1087. https://doi.org/10.1016/J.ENERGY.2015.01.061.

Zhang, Y., McKechnie, J., Cormier, D., Lyng, R., Mabee, W., Ogino, A., MacLean, H.L., 2010. Life cycle emissions and cost of producing electricity from coal, natural gas, and wood pellets in ontario, Canada. Environmental Science and Technology 44 (1), 538–544. https://doi.org/10.1021/es902555a.

Zhao, H., Wu, Q., Hu, S., Xu, H., Rasmussen, C.N., 2015. Review of energy storage system for wind power integration support. Applied Energy 137, 545–553. https://doi.org/10.1016/J.APENERGY.2014.04.103.

Zini, G., Tartarini, P., 2010. Wind-hydrogen energy stand-alone system with carbon storage: modeling and simulation. Renewable Energy 35 (11), 2461–2467. https://doi.org/10.1016/J.RENENE.2010.03.001.

Index

'*Note:* Page numbers followed by "f" indicate figures and "t" indicate tables.'

W

X

Z

Printed in the United States
By Bookmasters